W9-CCZ-814

TJ
790
D448
1993

Dempsey, Paul.

Small gas engine
repair.

DATE			
JUL 2 0 2002			

LIBRARY/LRC
OUACHITA TECHNICAL COLLEGE
P.O. BOX 816
MALVERN, ARKANSAS 72104

BAKER & TAYLOR BOOKS

Small
Gas Engine
Repair
second edition

Small
Gas Engine
Repair
second edition

Paul Dempsey

TAB Books
Division of McGraw-Hill, Inc.
Blue Ridge Summit, PA 17294-0850

SECOND EDITION
FIRST PRINTING

© 1993 by **TAB Books**.
TAB Books is a division of McGraw-Hill, Inc.

Printed in the United States of America. All rights reserved. The publisher takes no responsibility for the use of any of the materials or methods described in this book, nor for the products thereof.

The first edition of this book was published as *How to Troubleshoot and Repair Any Small Gas Engine*

Library of Congress Cataloging-in-Publication Data

Dempsey, Paul,
    Small gas engine repair / by Paul Dempsey. — 2nd ed.
      p.   cm.
    Includes index.
    ISBN 0-8306-4141-6   ISBN 0-8306-4142-4 (pbk.)
    1. Internal combustion engines, Spark ignition—Maintenance and repair. I. Title.
TJ790.D448 1993
621.43'4'0288—dc20   92-34047
                                        CIP

Acquisitions Editor: Kimberly Tabor
Editorial team: S. H. Mesner, Editor
                Susan Wahlman, Supervising Editor
                Joanne Slike, Executive Editor
                Jodi L. Taylor, Indexer
Production team: Katherine G. Brown, Director of Production
                Wendy L. Small, Layout
                Ollie Harmon, Typesetting
                Susan E. Hansford, Typesetting
                Kelly Christman, Proofreading
                Tara Ernst, Proofreading
                Marty Ehrlinger, Computer Artist
Design team: Jaclyn J. Boone, Designer
              Brian Allison, Associate Designer
Cover Design: Holberg Design, York, Pa.
Cover Photograph: Courtesy of Kohler Company, Wi.   TAB3

Contents

Introduction

The story about the technician who persuaded some giant, complicated machine to work by striking it with a hammer bears repeating. The owner objected to the $5000 invoice and asked for the charges to be itemized.

"Very well," said the technician, "One hammer blow: $5; knowing where to hit: $4995."

Almost anyone can repair a small (or large) engine. The actual work of making adjustments and changing parts is not difficult. The problem is knowing "where to hit."

The first edition of this book sold well (more than 100,000 copies), despite the fact that it was more concerned with hitting than aiming. That was a serious deficiency, because the problem facing mechanics is to discover what has gone wrong, what component has failed, or what adjustment has slipped. Diagnostics is the key.

Consequently, this second edition of *Small Gas Engine Repair* has been rethought and almost entirely rewritten. Only one chapter stands as it was. Most of the repair information has been retained, although it has been edited and presented more economically. Material dealing with obsolete engines has been deleted and material on new engines (such as the Briggs two-cycle) incorporated. But the stress is on *troubleshooting*.

This book is the first one on small engines to give troubleshooting skills the prominence they deserve. It is odd that the subject has received so little attention, because these skills are fundamental. A mechanic cannot listen to an engine run without analyzing the sound ("a little soft there—probably carburetor rich, maybe a dirty air filter"). Perhaps troubleshooting has been ignored because it has become transparent, like the grammar of our native language.

At any rate, writing the troubleshooting sections of this book involved tooth-pulling interviews with working mechanics (most of whom didn't know what the fuss was about), a search of factory literature (not usually helpful), and forced marches into the memory of years of engine work. I believe the techniques presented here will work for you.

The book is organized into six chapters, five of which deal with particular engine systems. You'll find a chapter on the fuel system, another on rewind starters (complex and troublesome enough to be considered a distinct system), another on charging systems, and so on.

The first chapter describes how to troubleshoot the engine as a complete, but malfunctioning, entity. If, for example, the engine refuses to start, turn to the "No Start" troubleshooting chart and trace the malfunction to a particular system. Once you have identified the system involved, turn to the appropriate chapter for information about that system.

Emphasis in the initial chapter is on distinguishing between ignition and carburetor-induced malfunctions. The ability to make this distinction is central to small engine work. Subsequent chapters contain troubleshooting information that, in most cases, enables you to trace the malfunction down to the component level.

In addition, these chapters include step-by-step repair procedures, illustrated by more than 180 line drawings and supported by specifications tables. The split between troubleshooting and repair depends upon the system. Charging systems are almost entirely a troubleshooting problem. Replacing a diode is not difficult, but you do have to know that the diode is bad. The chapter on rewind starters tilts toward repair. It is simple enough to determine what has gone wrong with one of these devices, but devilishly difficult to fix it.

I have been involved in small-engine work for many years. My first project was a Maytag washing-machine engine, which might give older readers some idea of the geologic time scale involved. This book reflects that experience, which has been confined for the most part to American-made utility and industrial engines. However, the techniques described here in the context of Tecumseh, Briggs & Stratton, Kohler, Onan, and Wisconsin apply, with minimal translation, to foreign engines as well.

1

Troubleshooting

The challenge and the frustration of small engine repair is *troubleshooting*. Automobile mechanics increasingly look to computers and electronic test instrumentation for their answers. Small-engine mechanics have what they were born with—five more-or-less fallible senses and native intelligence. They need to be smarter than the engine.

Before I can begin, I need to deal with some preliminary matters such as engine nomenclature, operating cycles, and ID information. Experienced mechanics can skip the next few pages, but readers who are new to small-engine work need to come to terms with this material. You cannot troubleshoot a machine that you do not understand. *Everyone*—regardless of skill level—should read the section on safety.

Nomenclature

Figure 1-1 illustrates and names the parts that make up a Wisconsin W1-145V single-cylinder, air-cooled, vertical-crankshaft, four-cycle engine. Unlike many other engines of this general type, it employs antifriction (ball-bearing) main bearings and a centrifugal governor. The flywheel doubles as a cooling fan.

Figure 1-2 shows an inexpensive two-cycle utility engine with the crankshaft deployed horizontally. This general type is known as a "bushing engine" because both ends of the connecting rod and the crankshaft run on plain bearings. These

1

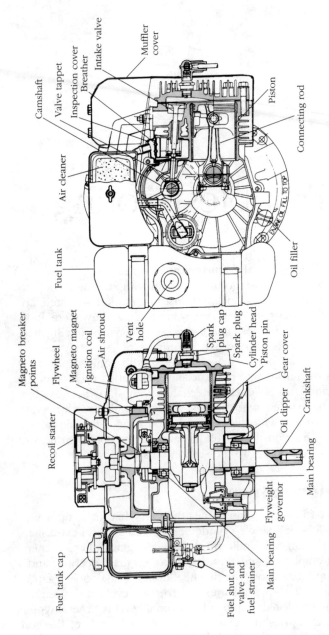

1-1 *Parts nomenclature for the Wisconsin Robin W1-145V Wisconsin Robin. Ball-bearing mains, generously sized parts, and high volume cooling system are evidence of quality.*

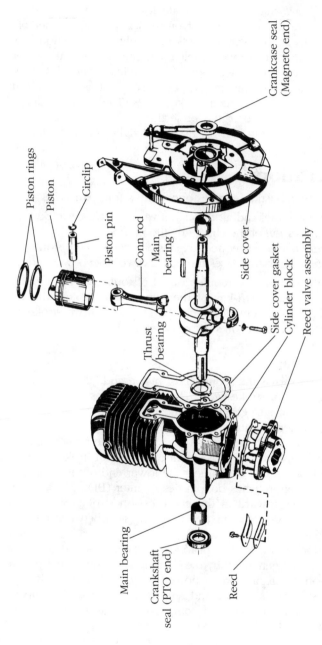

Piston rings

Piston

Circlip

Piston pin

Conn rod

Main bearing

Thrust bearing

Main bearing

Crankshaft seal (PTO end)

Reed

Crankcase seal (Magneto end)

Side cover

Side cover gasket

Cylinder block

Reed valve assembly

1-2 *The basic two-cycle engine, such as the Clinton shown here, is as simple as internal combustion gets. This particular engine seems to run best with a 0.016-inch point gap, rather than the 0.020-inch specified.*

bearings are marginally adequate for light-duty applications, but require large amounts of oil, typically on the order of 1 part oil to 24 or even 16 parts of gasoline. Antifriction bearings (ball or needle) would extend engine life and cut oil consumption by half. Other, more desirable features of the engine are the use of an integral cylinder head (cast as one with the barrel, or "jug") and reed intake valves. The piston is domed with the steep side of the projection adjacent to the transfer ports (visible about halfway up on the right side of the barrel).

Operation

Internal combustion engines operate in a four-part cycle:

- *Intake* of fuel and air through a valve or two-cycle port
- *Compression* of the charge by the piston
- *Ignition* and subsequent expansion against the piston
- *Exhaust* of burnt gases through a valve or two-cycle port.

Four-cycle engines require a total of four (two upward and two downward) strokes of the piston to complete the full cycle. Two-cycle engines telescope events into two strokes of the piston, or one crankshaft revolution.

Four-cycle

Figure 1-3 shows the sequence of piston and valve movement during the four-stroke cycle. The piston moves downward on the intake stroke, evacuating the cylinder above it. Air and fuel, impelled by atmospheric pressure, enter past the open intake valve to fill the void. The exhaust valve remains closed during the intake and subsequent compression stroke.

The intake valve closes near the lower end of the stroke, as the piston approaches bottom dead center (BDC). The piston now moves upward in the compression stroke. Since both valves are closed, the mixture is compressed between the piston top and the cylinder head.

Ignition occurs as the piston approaches top dead center (TDC). Force generated by the burning gases drives the piston down on the expansion, or power, stoke. The mass of the flywheel stores this energy and returns part of it during subsequent exhaust, intake, and compression strokes. Both valves remain closed during the expansion stroke.

The piston rounds BDC again and moves back up in the cylinder. The exhaust valve opens and the spent gases vent to

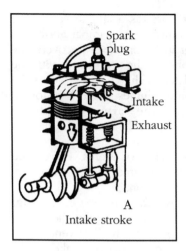

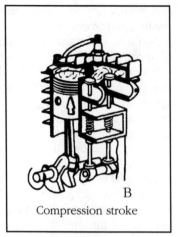

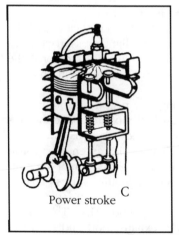

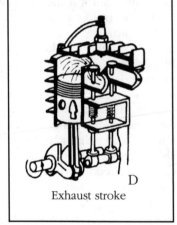

1-3 *The operating cycle common to all spark-ignition engines consists of four events—intake, compression, expansion, and exhaust. A four-stroke-cycle engine completes the cycle in two revolutions of the crankshaft.*

the atmosphere. The intake valve remains closed until the piston reaches TDC and a new cycle begins. Each cycle requires four strokes of the piston and two full crankshaft revolutions.

Two-cycle

The piston is the central element of a two-cycle engine. It functions both as a double-acting compressor and, in conjunction with ports in the cylinder bore, as a shuttle valve. The mixture is compressed in the combustion chamber above the piston and the crankcase under it. Piston travel uncovers one or more transfer ports that connect the crankcase with the cylinder bore and, somewhat later in the cycle, uncovers the exhaust ports. In addition, some engines use the piston to open an intake port connecting the carburetor with the crankcase. Two-cycle engines are mechanically simple and geometrically complex.

Figure 1-4 shows the basic sequence of events. The engine shown is similar to the one illustrated earlier in that it employs a reed intake valve and a deflector piston.

The upper left drawing shows the piston descending. Exhaust gases from the previous cycle flow over the piston crown and through the exhaust port, where they vent to the atmosphere. At the same time, the port labeled "intake" (most mechanics call it a transfer port) is open to connect the area above the piston with the crankcase. The crankcase and the fuel-air-oil mixture in it are subject to slight pressurization as the piston falls. Crankcase pressure is sufficient to force the fuel charge through the open intake (or transfer) port and into the cylinder bore.

In the upper right-hand drawing, the piston has progressed through BDC and is rising. The air-fuel mixture above the piston comes under increased compression, because both the transfer and exhaust ports are closed. The ascending piston also depressurizes the crankcase cavity. This partial vacuum causes the reed valve to spring open, admitting a fresh charge from the carburetor.

The spark plug fires as the piston rounds TDC. The piston moves down, first uncovering the exhaust port and then the transfer port.

But what forces the exhaust gases out? Two mechanisms are at work: Initially, these gases are at higher-than-atmospheric pressure and blow down naturally. Those gases that remain are scavenged by the incoming air-fuel charge. The incoming gas stream enters via the transfer port, strikes the piston deflector, and rebounds upward, driving the exhaust gases out before it.

This scheme, known as *crossflow scavenging*, has several drawbacks: Much of the exhaust remains in the cylinder, diluting the mixture; fairly large amounts of unburnt fuel are lost

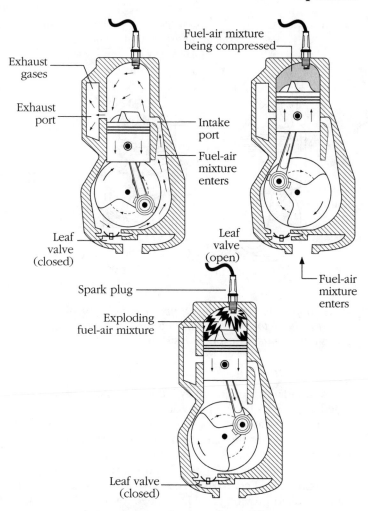

1-4 *Two-stroke-cycle engines compress the operating cycle into a single revolution of the crankshaft. In theory a two-cycle engine should develop twice the power of an equivalent four-cycle. In practice, the power advantage is somewhat less.*

to the exhaust; and the deflector makes the piston heavy and prone to distortion.

Loop-scavenging, shown in FIG. 1-5, represents a major improvement. The incoming charge enters through ports

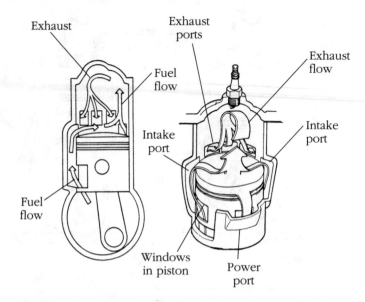

1-5 *A loop-scavenged engine uses angled transfer ports to impart a swirl to the incoming charge. The spinning charge scrubs the cylinder of exhaust residues.*

deployed radially around the cylinder circumference. Precisely angled exit ramps on the ports direct the gas streams upward, away from the flat-topped piston. The streams meet and form a vortex that drives the exhaust gases out ahead of it. Very little of the fresh charge escapes through the exhaust port.

Most utility and industrial engines use a reed valve between the carburetor and crankcase. The valve functions automatically and tends to provide good low-speed torque characteristics.

Tecumseh, Motobecane, and a few other manufacturers use a modification of the disk-type intake valve, once popular on racing motorcycles. These engines feed through a slot milled in the crankshaft, which is indexed to admit fuel on the piston upstroke. Another approach, favored by Homelite and European manufacturers, involves a third piston-controlled port. The port, located near the bottom of the cylinder barrel, connects the carburetor with the crankcase. It is uncovered near the top of the stroke and closed by the falling piston before the transfer port opens. The third-port system provides good piston lubrication, but imposes some limits on crankcase filling and engine power. Some of the charge squirts back into the carburetor as the port closes, which can make for a ragged idle.

Engine ID

You need the engine make, model number, and, in some cases, the serial number to order parts. Alphanumeric information is usually stamped on the blower housing or on the block flange (FIG. 1-6). Engine manufacturers usually identify their products with a decal, supplemented by a characteristic color, keyed to the engine model. With a little practice, you will be able to spot the products of the various makers at a glance.

Safety

The chief hazard associated with small-engine repair work is gasoline, especially when it has escaped the confines of the fuel tank. Work in a well-ventilated area, away from ignition sources. If gasoline has spilled onto the engine, allow it to evaporate before proceeding with the work. *Do not* attempt to start a wet engine. *Never* refuel or open a fuel line on a hot engine. Some blends of gasoline ignite at temperatures slightly above the boiling point of water. Your local gasoline distributor can verify that statement.

Small-engine fires usually occur during or just after starting, while the mechanic is standing nearby or leaning over the engine. I have been affected by the legal and medical fallout from several small-engine accidents and hope never to hear of one again. Do not become so involved in doing the next thing that you lose sight of your own safety.

A second hazard is asbestos, used in high-temperature gaskets and apparently in the flywheel brake shoes fitted to some rotary lawnmower engines. Asbestos fibers are carcinogenic when inhaled or ingested. One way to deal with the problem—although I cannot guarantee its absolute safety—is to lightly grease gaskets before removal. Scrape the residue with a razor blade, disposing of the fragments in a sealed trash container. Results are less than perfect—some discoloration remains, but this is far safer than removing gasket shards with a wire wheel or wire brush.

Use conventional solvents, such as kerosene or Gunk. The latter product, available in premixed or concentrated form from automotive parts houses, rinses off with water. It does a

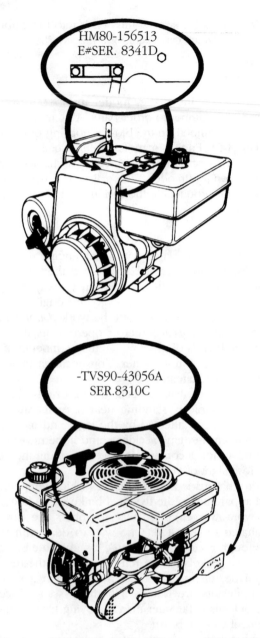

1-6 *Engine ID information is usually stamped on the blower housing or flange area. Starter motors, carburetors, and other subassemblies also carry an identifying number.* Tecumseh Products Co.

marvelous job on grease-encrusted parts, but because of the water rinse, Gunk is not compatible with electrical parts.

Finally come the hazards associated with rotating machinery. Keep hands and fingers clear of V-belts and other moving parts. Do not work on the power takeoff end of any engine without defeating the ignition by disconnecting the spark plugs. In some cases, merely disconnecting the lead is not enough, since the wire "remembers" where it has been and tends to float back into proximity with the spark plug. Bend the wire or wedge it into the cylinder finning.

Warnings and **Cautions** in this book follow military specification practice. **Warning** means risk of personal injury; **Caution** means risk of equipment damage.

Troubleshooting

An engine should run if it has fuel, 60 psi of cranking compression, and a spark synchronized with piston movement. The fuel system provides the combustible mixture, the engine mechanical system inducts and compresses that mixture, and the ignition system generates the timed voltage pulse that ignites the mixture.

The first step in troubleshooting is to determine which of these three systems has malfunctioned. Other systems, such as the charging system or electric starting system, are marginal to our purpose, since the engine can run without them. The logic trees in FIGS. 1-7 through 1-13 should help you identify the failed system. Each chapter in this book is keyed to a particular system and prefaced with troubleshooting information. Once you have identified the failed system, turn to the appropriate chapter and troubleshoot the system to the component level.

This approach to troubleshooting applies to all small utility engines, regardless of make or type. It works more than 90 percent of the time. But no experienced mechanic believes that all answers can be found in a book. A hard core of problems elude systematic detection. Some malfunctions are just too obvious to register—the cracked spark plug insulator or the loose wire that your 12-year-old spots immediately. Other malfunctions are of a more serious and frustrating nature. Troubleshooting techniques are based on the idea that each problem generates a single, easily recognizable symptom. That is not quite true. Ignition maladies can easily be confused with

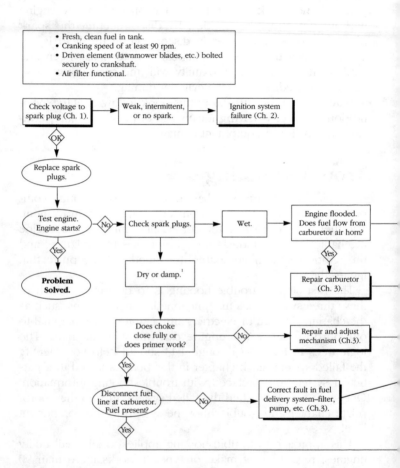

1-7 *Troubleshooting procedure—engine does not start.*

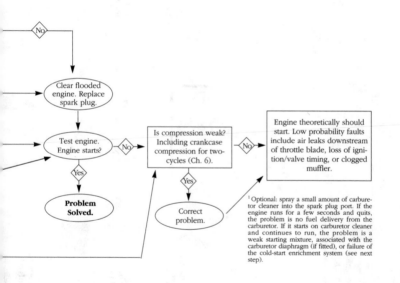

No

Clear flooded engine. Replace spark plug.

Test engine. Engine starts? No

Yes

Problem Solved.

Is compression weak? Including crankcase compression for two-cycles (Ch. 6). No

Yes

Correct problem.

Engine theoretically should start. Low probability faults include air leaks downstream of throttle blade, loss of ignition/valve timing, or clogged muffler.

[1] Optional: spray a small amount of carburetor cleaner into the spark plug port. If the engine runs for a few seconds and quits, the problem is no fuel delivery from the carburetor. If it starts on carburetor cleaner and continues to run, the problem is a weak starting mixture, associated with the carburetor diaphragm (if fitted), or failure of the cold-start enrichment system (see next step).

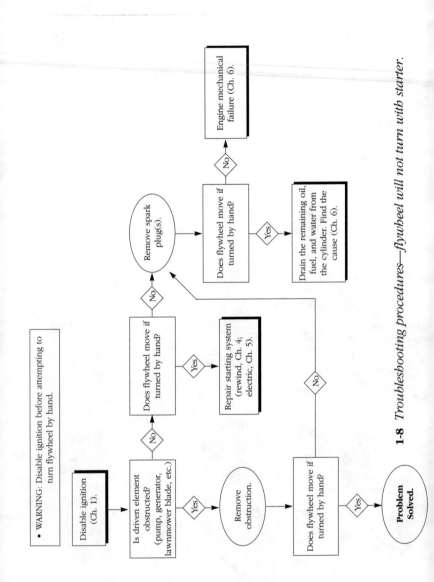

1-8 *Troubleshooting procedures—flywheel will not turn with starter.*

carburetor malfunctions; a loose blade on a rotary lawnmower can sound like a rod knock.

Troubleshooting techniques of whatever kind also assume that problems come in ones, like ducks in a row. In the real world problems can come in troops and legions. Everything that can go wrong, does. Admittedly this condition is rare and limited to old, well-worn engines or engines that have not been started for years. But it does happen. You can only go through each system in turn, correcting malfunctions as they are encountered and trusting that the problems are finite.

Materials

You need the following parts and supplies:

- A known-good spark plug, gapped to specification. Small engines, particularly those with marginal magneto ignitions, can be very fussy about spark plugs. A new spark plug should not be considered functional until it is tested in an engine.
- A freshly charged battery for electric-start models.
- Clean, fresh gasoline (or premixed fuel for two-cycle engines).
- A container to catch spilled fuel.
- Motor oil for topping off the crankcase of four-cycle engines.
- Standard shop tools plus a spark tester and a flywheel-removal tool. You might also need a tachometer, although you can probably get by without one if you do not attempt to make governor adjustments or change the idle rpm setting. Many small engines are designed to idle quite fast—in excess of 2000 rpm.

  Caution: Very low idle speeds can result in catastrophic failure when the throttle is suddenly opened.

  Warning: The safe upper rpm limit is fixed by the governor; if this mechanism is defeated, the engine might explode in a shower of mechanical fragments.

- Wynn's Carburetor Cleaner or the equivalent in aerosol form.
- A supply of clean rags.

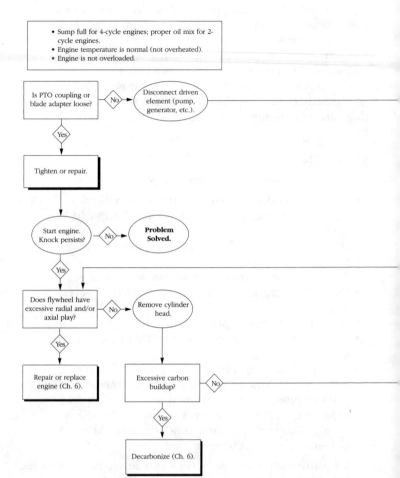

- Sump full for 4-cycle engines; proper oil mix for 2-cycle engines.
- Engine temperature is normal (not overheated).
- Engine is not overloaded.

Is PTO coupling or blade adapter loose? —No→ Disconnect driven element (pump, generator, etc.).

↓Yes

Tighten or repair.

Start engine. Knock persists? —No→ **Problem Solved.**

↓Yes

Does flywheel have excessive radial and/or axial play? —No→ Remove cylinder head.

↓Yes

Repair or replace engine (Ch. 6).

Excessive carbon buildup? —No→

↓Yes

Decarbonize (Ch. 6).

1-9 *Troubleshooting procedure—engine runs, but knocks.*

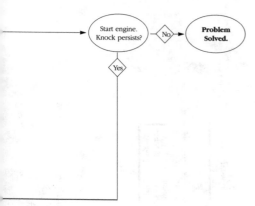

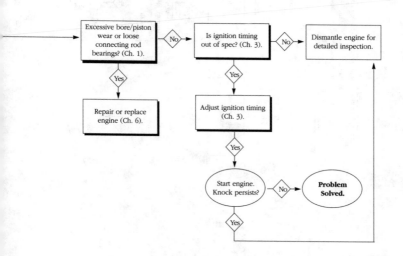

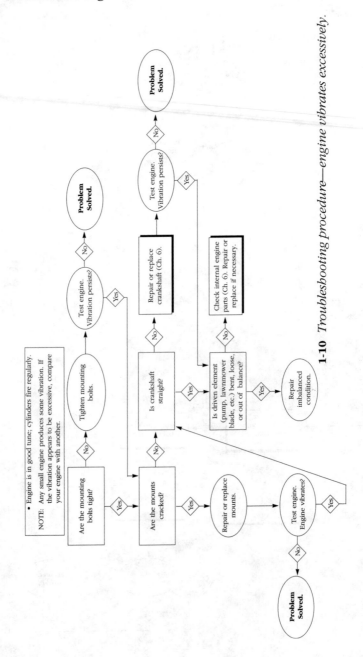

1-10 *Troubleshooting procedure—engine vibrates excessively.*

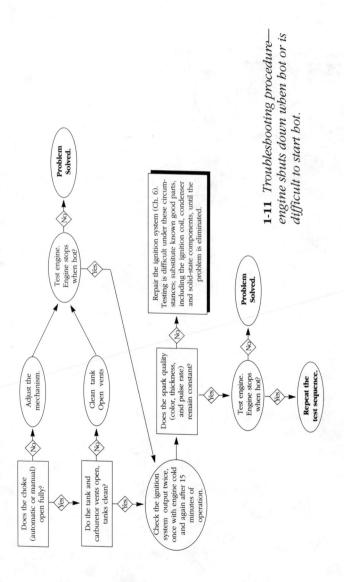

Does the choke (automatic or manual) open fully?

Yes / No

Adjust the mechanism.

Do the tank and carburetor vents open, tanks clean?

Yes / No

Clean tank Open vents

Test engine. Engine stops when hot?

Yes / No

Problem Solved.

Check the ignition system output twice, once with engine cold and again after 15 minutes of operation.

Does the spark quality (color, thickness, and pulse rate) remain constant?

Yes / No

Repair the ignition system (Ch. 6). Testing is difficult under these circumstances; substitute known good parts, including the ignition coil, condenser and solid-state components, until the problem is eliminated.

Test engine. Engine stops when hot?

Yes / No

Problem Solved.

Repeat the test sequence.

1-11 Troubleshooting procedure— engine shuts down when hot or is difficult to start hot.

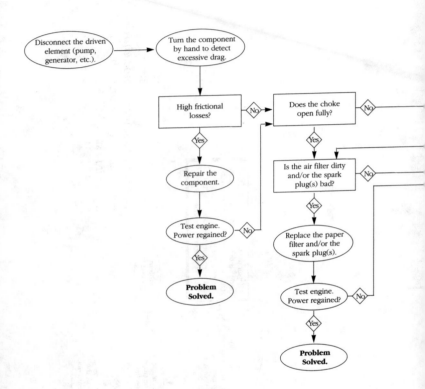

1-12 *Troubleshooting procedure—engine lacks power, without obvious misfiring.*

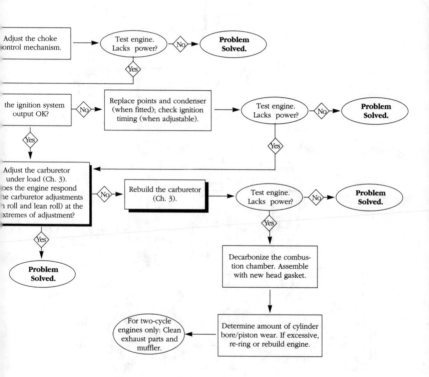

Adjust the choke control mechanism. → Test engine. Lacks power? → No → **Problem Solved.**

Yes ↓

the ignition system output OK? → No → Replace points and condenser (when fitted); check ignition timing (when adjustable). → Test engine. Lacks power? → No → **Problem Solved.**

Yes ↓

Adjust the carburetor under load (Ch. 3). Does the engine respond the carburetor adjustments (rich roll and lean roll) at the extremes of adjustment? → No → Rebuild the carburetor (Ch. 3). → Test engine. Lacks power? → No → **Problem Solved.**

Yes ↓

Problem Solved.

Yes ↓

Decarbonize the combustion chamber. Assemble with new head gasket.

↓

For two-cycle engines only: Clean exhaust parts and muffler. ← Determine amount of cylinder bore/piston wear. If excessive, re-ring or rebuild engine.

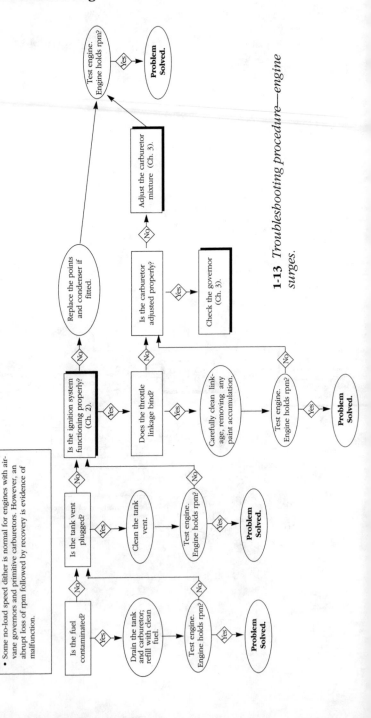

1-13 *Troubleshooting procedure—engine surges.*

Preliminaries

Begin by checking the crankcase oil level on four-cycle engines. No visible oil, or black, gritty oil that has been carburized by heat mean bearing damage. You can get an idea of the extent of the damage from the amount of bearing play at the flywheel. Remove the shroud and alternately pull and push on the flywheel. Some axial movement—on the order of 0.006 inch, or about the thickness of a piece of bond paper—is normal. Excessive play is accompanied by an audible click as the crankshaft thrust surface contacts its bearing. Attempt to move the flywheel side-to-side. You should just be able to sense radial movement, which on a new engine is about 0.002 inch. Turn the flywheel to bring the piston to the top of its stroke. It might be helpful to remove the spark plug and track piston movement with the aid of a flashlight or a screwdriver inserted into the spark plug port. With the piston at top dead center, rock the flywheel a few degrees in both directions. Piston movement should track flywheel movement almost exactly. Some "slop" is normal and represents the big-end bearing clearance, but if the piston remains stationary through two or three degrees of flywheel movement, the connecting rod and/or crankshaft pin are worn.

Generally, two-cycle engines and most older four-strokes can be tested with a compression gauge. With the ignition OFF, and the throttle and choke wide open, the typical engine generates 80 to 100 psi (pounds per square inch) at normal cranking speeds. At least 60 psi is necessary for starting.

Caution: This and other test procedures call for cranking the engine with the spark plug wire disconnected. Make certain the ignition is OFF. If the engine does not have an ignition switch, ground the spark plug cable to the engine block. Ignition systems—especially transistorized systems—can self-destruct if energized with the output ungrounded.

Modern four-cycle engines typically include a compression release, which unseats the exhaust valve during cranking or when the engine is turning at very low rpm. Most of these mechanisms can be defeated by cranking the engine backwards with a rope. But the effort is hardly worthwhile; with a little experience, you can feel engine compression as resistance on the starter cord. With more experience, you can feel crankcase compression on a two-cycle engine, the absence of which makes starting impossible. A possible early indication of two-cycle loss of crankcase compression is refusal of the engine to run hot without the choke engaged. Air enters past the crankcase seals and leans out the mixture.

Some low-performance two-cycle engines tend to carbon over their exhaust ports, a condition that costs power. Remove the muffler (which also can become carbon-clogged) and check that ports are clear. To clean, lower the piston below port level and scrape the carbon with a brass or copper tool (FIG. 1-14). Remove the spark plug, ground the ignition, and spin the engine to remove loose carbon flakes.

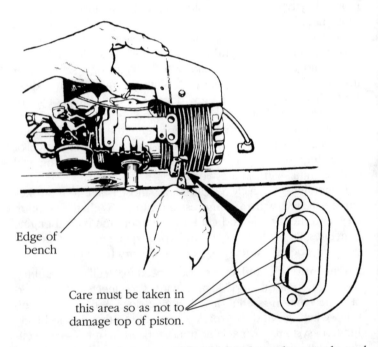

Edge of
bench

Care must be taken in this area so as not to damage top of piston.

1-14 *Two-cycle exhaust ports should be cleaned routinely and whenever there is a loss of power at wide throttle angles.* Tecumseh Products Co

Check the fuel supply and quality early in the troubleshooting process. The open shelf life of gasoline is about six months. Summer-grade gasoline can cause hard starting in subzero weather. If you have any question about the fuel, drain the tank and carburetor and refill with fresh gasoline. Consult the owner's manual to determine the correct proportion of oil and gasoline for the two-cycle engine in question. Too little oil accelerates cylinder bore wear and can result in piston seizure or crankpin damage; too much oil makes starting difficult, lowers the octane rating of the fuel, and shortens spark plug life. Oil and fuel must be thoroughly mixed by agitating the container prior to refueling.

Warning: Conduct refueling operations in a well-ventilated area—preferably outdoors—and away from possible ignition sources. The engine must be cold. Wipe up spilled fuel and allow time for fuel vapor to dissipate before proceeding with repairs.

If the engine is fitted with a choke valve, make sure it fully closes. Most engines do not start cold unless the choke blade completely blocks the carburetor air horn. Adjust the linkage as necessary.

Check the air filter. Excessively dirty air filters cost power and, in extreme cases, can make starting difficult or impossible. Polyurethane foam filter elements can be washed in kerosene and lightly oiled: a teaspoon of oil is sufficient. These filters should be reoiled periodically, because the oil tends to migrate to the downstream side of the element. Clean oil bath filters in solvent, wipe out the filter housing, and fill to the indicated level. Overoiling causes the engine to smoke. Paper filters should be replaced.

Caution: Pleated paper filters lose porosity if wetted with solvents or water.

Most engine faults—hard starting, misfiring, hot shutdowns—originate in the ignition system, particularly if that system consists of a conventional (point and condenser) magneto. Solid-state systems are more reliable and generally shut down completely on failure.

Ignition systems are tested by observing the spark as it jumps an air gap between the ignition cable and an engine ground. Figure 1-15 shows an ignition tester made from a new RJ-8 or RCJ-8 spark plug. The ground electrode is removed to give a gap of about 0.13 inch, which simulates conditions inside the engine. The fuel line slipped over the spark plug threads acts as a shield to make the spark more visible.

Conventional (i.e., battery and coil or magneto) systems should deliver a thick blue spark that snaps audibly on discharge. A red or branching white spark indicates low system voltage. Capacitive-discharge ignitions (CDIs) are hot enough

1-15 Connect the tester to the high-tension lead as shown; the smaller alligator clip attaches to a cylinder-head fin. Spin the flywheel at normal cranking speeds (about 150 rpm) for all applications except Briggs & Stratton Magnatron ignition systems. These solid-state systems require 350 rpm or so for excitation and do not produce the blue spark characteristic of healthy magnetos. (Courtesy Kohler.) While this Kohler-designed tester is perfectly adequate, you might prefer the new Briggs & Stratton type. PN 19368 incorporates a 0.166-inch gap shielded behind a window to reduce shock and fire hazard.

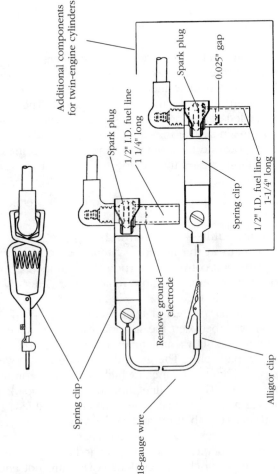

Additional components for twin-engine cylinders

Spark plug

1/2" I.D. fuel line 1 1/4" long

0.025" gap

Spark plug

Spring clip

1/2" I.D. fuel line 1-1/4" long

Remove ground electrode

Spring clip

18-gauge wire

Alligtor clip

to burn paint: solid-state systems without a charging capacitor deliver what appear to be puny sparks, but function normally on the engine. Any solid-state ignition that pumps out enough voltage to jump the 0.13-inch spark gap should be considered okay. The spark should pulse, once every revolution for most engines and once every two revolutions for the more-sophisticated four-strokes. Missed pulses indicate a system malfunction.

Caution: Use an appropriate tool for testing the spark. Do not simply disconnect the cable and hold its end near an engine ground. This technique poses the risk of electroshock and—if the air gap is too large—can destroy the coil. Modern ignition systems deliver 50,000 V, open-circuit.

A spark plug can look normal and fire outside of the engine, but fail to fire under compression. Eliminate uncertainty by substituting a known-good spark plug early in the troubleshooting process.

Carburetor-related faults are a distant second to ignition in frequency. Diaphragm-type carburetors do, however, give problems, especially if stored wet. Change out the diaphragm whenever the engine shows a reluctance to start.

Do not continue to crank a dead engine. The exercise is usually a waste of time and floods the combustion chamber with raw fuel, further complicating the situation.

Clear flooded chambers with compressed air applied through the spark plug port (which can cause the engine to "motor"), or simply wait a few minutes for most of the fuel to evaporate and substitute a dry spark plug. A spark-output tester, such as B & S PN 19368, connected in series with the spark plug, can boost voltage.

Flooding in two-cycle engines can be very severe, because liquid fuel can collect in the crankcase. Some early engines were fitted with crankcase drain plugs for this eventuality; the best you can do with a modern engine is wait for the surplus to evaporate.

Oil flooding occurs after very long periods of cranking four-cycle engines or if, for some reason, the engine has been inverted in the piston-down position. Carburetor cleaner, compressed air, and a supply of dry spark plugs will eventually solve the problem.

I should also mention that electric starters have very limited duty cycles. If you can crank the engine by hand, do so.

Finally, do not make frivolous carburetor or ignition-timing adjustments when attempting to start a balky engine. These adjustments do not change over the short term; if the engine ran before, it should run now.

2

Ignition system

Most small engine malfunctions—starting difficulties, misfiring, and loss of power without obvious misfire—originate in the ignition system. Check this system first.

System recognition

The first step in troubleshooting is to identify the system. Until recently, most small engines were fitted with flywheel magnetos. In nearly all applications, the breaker points live under the flywheel and drive from a crankshaft cam. More sophisticated magnetos locate the points outside of the flywheel, driving them at half engine speed from the camshaft. Some few magnetos, such as the old Fairbanks-Morse, are entirely self-contained units mounted at the end of the camshaft. In all examples, magnets energize the ignition coil and breaker points perform the switching function.

From a mechanic's point of view, battery-and-coil ignition can be thought of as a friendlier, more accessible cousin of the magneto. Both systems employ an ignition coil, a point set, and a condenser. Secondary circuits are identical, but the B & C system draws its primary voltage from a storage battery, which eliminates the uncertainties associated with spinning flywheel magnets.

CDI (capacitive discharge ignition) systems show wide variation in architecture. Most include an under-flywheel trigger coil and a discrete ignition coil. Flywheel magnets provide the energy and transistors take the place of breaker points.

Caution: CDI systems are fragile and can be damaged by routine troubleshooting procedures. Observe the precautions under "CDI Systems" below.

Troubleshooting the system

Figure 2-1 describes how to troubleshoot any ignition system, regardless of type or manufacturer. A general approach is possible, since all ignition systems produce—or should produce—a pulse of high-voltage electrical energy timed to occur as the piston nears top dead center.

Replacement parts

Without test instruments more sophisticated than a multimeter, you must put your trust in the parts you have replaced. New or known-good parts provide you with a moving baseline of discarded theories. Things get confusing when replacement parts create additional malfunctions.

Replace sacrificial parts—spark plugs, point sets, and condensers—with new OEM (original equipment manufacturer) parts. Ignition coils, CDI modules, and pulse transformers cannibalized from other engines can save money, but make certain that components function and carry the same part numbers as the originals. Bolt-on capability does not ensure electrical compatibility.

Spark plug

Replace the spark plug early on in the process. Use a new and tested plug of the type recommended for the application and gapped to specification. Start the plug by hand, turning it through three full revolutions before applying a wrench.

Spark output test

Remove the spark plug and connect the high-tension cable to a spark-output tester, as shown in FIG. 1-15. Ground the tool and spin the flywheel. As mentioned in Chapter 1, conventional systems should deliver a thick, electric-blue spark with perhaps a few reddish whiskers. Solid-state systems might not deliver with such authority, but however red and spindly, the spark should jump the tester gap.

Warning: All testers pose a fire hazard. Make certain no fuel or fuel vapor is present.

Note: Briggs & Stratton Magnetron systems are slow to come to life and require a cranking speed of at least 350 rpm.

Failure to fire is usually caused by a single, easy-to-correct malfunction. Begin with a careful scrutiny of the external primary circuit. Look for shorts (burned or frayed insulation), opens (broken wire, corroded connections), and failed interlocks.

Testing interlocks

An interlock is an automatic switch in the primary circuit that shuts down the ignition under unsafe operating conditions. Function varies with the application, and the more elaborate circuits include a logic module. But from a mechanic's point of view, interlocks fall into two categories: those that are normally open (NO), and those that are normally closed (NC). NO switches shunt primary current to ground when activated (FIGS. 2-2, 2-3); NC switches open the primary circuit. In either case, ignition is denied.

To test switch function, disconnect the associated wiring and test switch continuity with an ohmmeter. Resistance should be essentially zero with contacts closed and infinity with contacts open.

Caution: Do not operate the engine with interlock leads disconnected or grounded. System damage can result.

If the problem persists, turn your attention to the spark generator itself. In nearly all instances, you will need to remove the engine cooling shroud and flywheel.

Flywheel removal and installation

Flywheels are secured to the tapered end of the crankshaft by a key, a nut, and (usually) a lockwasher. Most engines turn clockwise when viewed from the flywheel end and use a conventional right-hand flywheel nut thread. Engines with opposite rotation can employ a left-hand thread.

Note: Early Briggs & Stratton engines, built before the days of rewind starters and now collector's items, had left-handed flywheel nuts.

- Ignition switch ON
- System interlocks disengaged
- Battery charged (when present)
- Known–good spark plug installed
- Spark equality tested weak following the procedure described in Chapter 2.

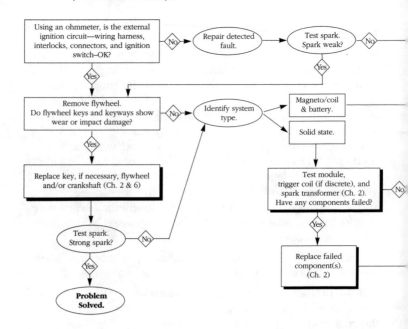

2-1 (A) *Ignition system troubleshooting—weak, erratic, or no spark.*

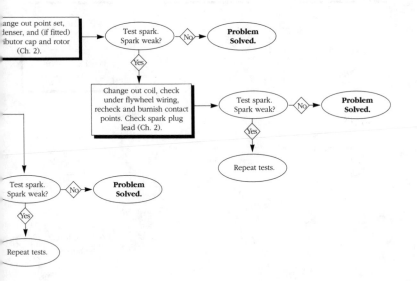

Problem
Solved.

...ange out point set,
...denser, and (if fitted)
...ibutor cap and rotor
(Ch. 2).

Test spark.
Spark weak? No **Problem
Solved.**

Yes

Change out coil, check
under flywheel wiring,
recheck and burnish contact
points. Check spark plug
lead (Ch. 2).

Test spark.
Spark weak? No **Problem
Solved.**

Yes

Repeat tests.

Test spark.
Spark weak? No **Problem
Solved.**

Yes

Repeat tests.

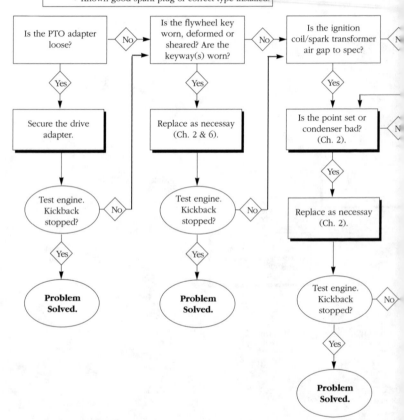

- Spark output OK.
- Cooling system OK—shrouds in place, fins clean.
- Known-good spark plug of correct type installed.

Is the PTO adapter loose? — No →

Is the flywheel key worn, deformed or sheared? Are the keyway(s) worn? — No →

Is the ignition coil/spark transformer air gap to spec? — N

Yes ↓

Secure the drive adapter.

Yes ↓

Replace as necessay (Ch. 2 & 6).

Yes ↓

Is the point set or condenser bad? (Ch. 2). — N

↓

Test engine. Kickback stopped? — No

↓

Test engine. Kickback stopped? — No

Yes ↓

Replace as necessay (Ch. 2).

Yes ↓

Problem Solved.

Yes ↓

Problem Solved.

↓

Test engine. Kickback stopped? — No

Yes ↓

Problem Solved.

2-1 (B) *Ignition system troubleshooting kickback during cranking* .

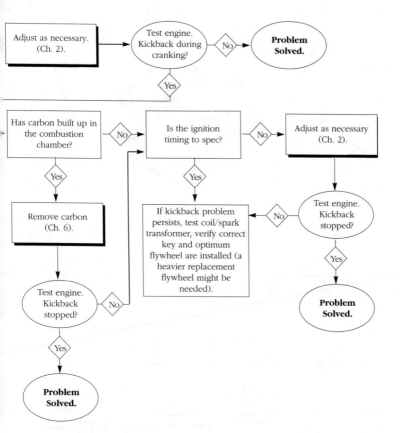

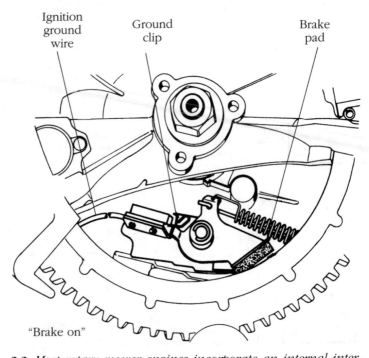

Ignition ground wire

Ground clip

Brake pad

"Brake on"

2-2 *Most rotary mower engines incorporate an internal interlock that grounds the primary ignition circuit when the brake is engaged. Test with an ohmmeter or by disconnecting the ground lead.*

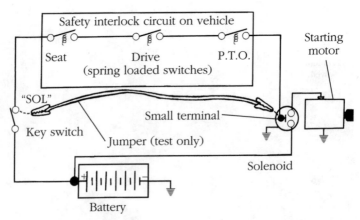

Safety interlock circuit on vehicle

Seat   Drive   P.T.O.
(spring loaded switches)

Starting motor

"SOL"

Small terminal

Key switch

Jumper (test only)

Solenoid

Battery

2-3 *External interlocks, associated with driven equipment, can be tested with a jumper.*

Remove the retaining nut with the appropriate metric or inch-standard socket. The light-metal retainer nut/starter clutch on most single-cylinder Briggs & Stratton engines requires a special tool, PN 19161.

Secure the flywheel with a strap wrench (FIG. 2-4). Rotary lawnmower blades can be blocked with a short piece of two-by-four.

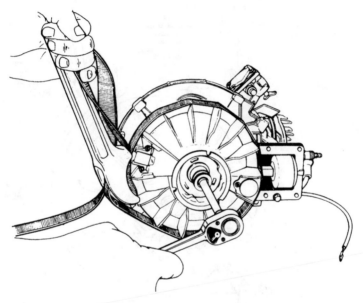

2-4 *Use a strap wrench to restrain the flywheel against retainer nut torque.*

Caution: Do not attempt to lock the flywheel by using one of the vanes as a purchase point. The vane may snap off, throwing the flywheel out of balance.

Most retainer nuts are backed up with a lockwasher. Bellville spring washers install with the concave (dished) side of the washer next to flywheel.

There are at least four ways to separate the flywheel from the crankshaft taper and key. In order of ascending violence, they are:

• Heavy cast-iron wheels can usually be jimmied off with two screwdrivers inserted between the back side of the wheel and the crankcase.

- Most aluminum flywheels should be removed with a special puller, available from the engine distributor or (for popular models) from automotive parts houses. The puller threads into predrilled holes in the flywheel hub and reacts against the crankshaft end (FIG. 2-5).

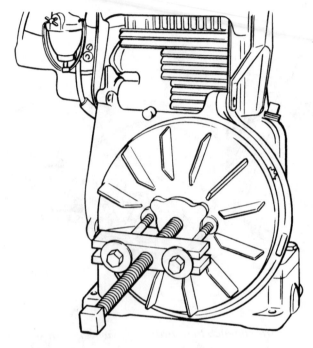

2-5 *A flywheel puller can be fabricated or ordered from the factory (Kohler shown). Small European engines can present a problem because the puller threads into a counterbore in the flywheel hub. If you cannot obtain a puller from the usual sources, try bicycle dealers. A similar tool is used to disassemble cotterless cranks.*

Caution: A conventional gear puller that attaches to the flywheel rim will distort the flywheel and can break it.

- Some flywheels can legitimately (i.e., with the factory's blessing) be shocked off with a knocker (FIG. 2-6). Tecumseh supplies three of these tools: PN 670103 for right-handed ⁷⁄₁₆-inch 20-TPI (threads per inch) shafts;

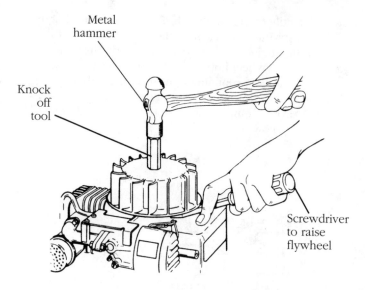

2-6 *Millions of flywheels have been successfully knocked off, but the procedure entails risk. Thread the knocker over the flywheel, back off a few turns, and hit squarely.*

PN 670118 for the left-hand variant of the same thread; and PN 670169 for standard ⁷⁄₁₆-inch, 20-TPI shafts. Thread the knocker down until it seats against the fly-wheel and back-off two or three turns. Insert a large screwdriver under the flywheel (not the stator plate) and pry up while giving the knocker a sharp rap with a hammer. Hit square and as hard as necessary to shock the wheel off. A glancing blow can break the crankshaft, most of which are made of cast iron are and quite brittle.

Warning: Wear eye protection when using a hammer on steel.

• Knockers are brutal tools at best, and should not be used on engines with antifriction (ball, roller, roller taper, or needle) main bearings. In addition to bearing damage, the impact can displace the crankshaft, forcing it into rubbing contact with the crankcase (FIG. 2-7).

• There are times when even more severe measures are in order—the retaining nut has been overtorqued, the key has seared and jammed, or corrosion has welded the wheel solidly to the crankshaft. Puller holes strip out and the knocker threatens to drive the crankshaft out of the end of the engine. In these extreme cases, it is permissible to heat the flywheel hub with a propane torch. Do not overdo it, and concentrate the heat on the hub, moving the flame constantly. At all cost avoid heating flywheel areas near electronic components. The flywheel should expand enough to allow removal.

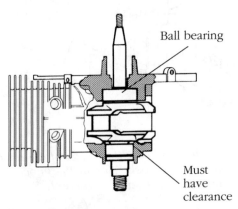

Ball bearing

Must have clearance

2-7 *Knocker-induced crankshaft dislocation can be corrected with a blow from a rawhide mallet on the PTO end of the shaft. Engines affected employ antifriction mains pressed into the crankless halves. Tecumseh two-stroke shown.*

Flywheel key and keyway

Inspect the flywheel hub for cracks, which generally radiate outward from the keyway (FIG. 2-8). Cracks too small to be seen by the naked eye can be detected with the ZyGlo or equivalent penetrant dye processes. Most automotive machinists can provide this service at a nominal cost.

Warning: Damaged flywheels represent a major hazard for anyone in the vicinity of the engine.

Remove the flywheel key from the crankshaft stub. Stubborn keys can be extracted with side-cutting diagonal pliers. Replace the key if worn or damaged (FIG. 2-9). Using a new key as a gauge, check both flywheel and crankshaft keyways for wallow. Small dislocations of flywheel rim magnets can make starting difficult. Unfortunately, there appears to be no

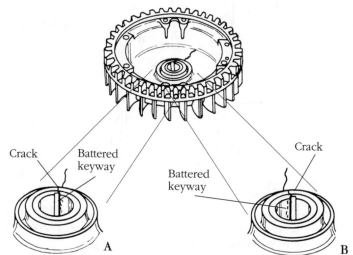

2-8 *A cracked flywheel is bad news, but there is some consolation in knowing how it happened. It is assumed that the normal direction of engine rotation is clockwise when viewed from the flywheel end. A crack on the leading edge of the keyway (A) means that the crankshaft was overspeeding the flywheel. This can occur only if the flywheel retaining nut is loose. A crack on the trailing edge (B) suggests that the crankshaft stopped or suddenly slowed, allowing the flywheel to overrun it. Expect to find collateral damage, including a bent or twisted crankshaft.*

cost-effective way to restore keyway integrity; either or both parts must be replaced.

Alnico and the newer ceramic flywheel magnets do not appreciably weaken in normal service. However, problems have been reported with Repco pin-type magnets, which press into drilled holes in the flywheel rim. As a rule, a generating magnet that attracts a loosely held screwdriver through an air gap of three-quarters of an inch can be considered good. Smaller magnets used to energize CDI trigger coils need not be that powerful.

You need not completely assemble the engine to test system output. Replace the key (FIG. 2-10 shows correct lays for

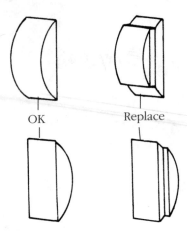

OK   Replace

2-9 *It is good practice to replace the key whenever the flywheel is lifted. Do not substitute a steel key for the Briggs & Stratton aluminum type illustrated.*

plane and Woodruf keys) and slip the flywheel over the crankshaft. With the spark plug removed and system output grounded through an air gap, spin the flywheel by hand. If all is right, a spark will be generated. Tighten the retainer nut to specification.

Warning: Be careful not to cut your fingers on the governor air vane or other obstructions near the flywheel rim.

Loss of timing

At some point in the troubleshooting process (usually near the point of physical collapse), you might begin to suspect that the ignition timing is at fault. Avoid such speculations; many engines (Briggs & Stratton, Yamaha, etc.) have fixed timing; others are adjustable over a narrow range, defined by point-gap variations. An out-of-time condition severe enough to affect starting will usually defeat the magneto (but not necessarily CDI systems). Other engines have slotted stator plates that allow fairly wide variations in timing, but upon examination you usually find the hold-down bolts rusted into place. Nor does timing drift much on small engines; if the engine ran since disassembly, it should still be in time.

Yet on occasion, an engine spontaneously "jumps time" without losing adjustment. You can usually sense the loss of

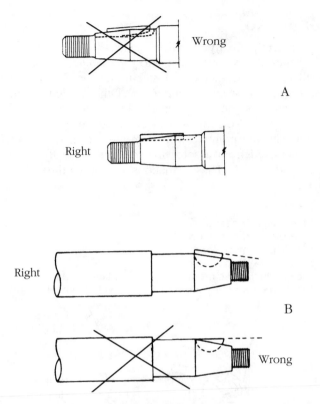

2-10 *Most crankshaft keyways are failsafe; Kohler (A) and OMC (B) keys can be improperly installed.*

time during cranking; the starter cord pulls hard and might "bite" your hand as the engine tries to run backwards.

Spontaneous loss of time appears to be limited to systems that receive their firing impulses from the flywheel. The usual culprits are a loose flywheel retaining nut, a sheared or deformed key, or "wallowed" keyways. A loose PTO (power takeoff) adapter, which allows the driven element to move relative to the crankshaft, can have the same effect on engines with skimpy flywheels. Some rotary lawnmowers will not start unless the blade is secure, which must be one of the few instances when manufacturing economies have actually promoted safety.

If you must retime the engine, follow the instructions under "Timing."

Magneto

Until recently, nearly all small engines were fired by magnetos. Figure 2-11 shows a magneto for a single-cylinder engine. In this example, all parts cluster under the flywheel, which has magnets cast into its inner rim. Other designs mount the coil outside of the flywheel, in which case the rim magnets face outward. You might also encounter magnetos with remotely situated points driven from the camshaft, and magnetos energized from a rotor housed under the flywheel, but independent of it. These variations are shown throughout this chapter.

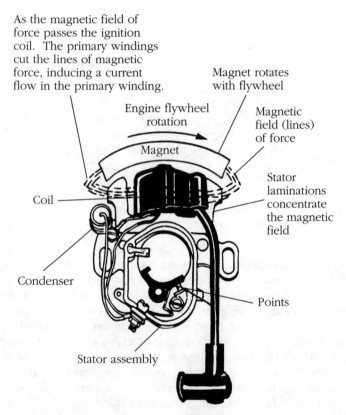

As the magnetic field of force passes the ignition coil. The primary windings cut the lines of magnetic force, inducing a current flow in the primary winding.

Magnet rotates with flywheel

Engine flywheel rotation

Magnetic field (lines) of force

Magnet

Coil

Stator laminations concentrate the magnetic field

Condenser

Points

Stator assembly

2-11 *Magneto parts arrangement and nomenclature. The point cam is omitted for clarity.*

The ignition coil acts as the interface between the primary and secondary circuits. The primary winding consists of about 200 turns of copper wire, wrapped over the stator laminations, (as shown in FIG. 2-12). One end of the winding grounds to the engine; the other end connects to the movable point arm. The secondary winding is made up of some 10,000 turns of hair-fine copper wire, wound on top of the primary but insulated from it. One end of the secondary is grounded and the free end terminates at the spark plug cable.

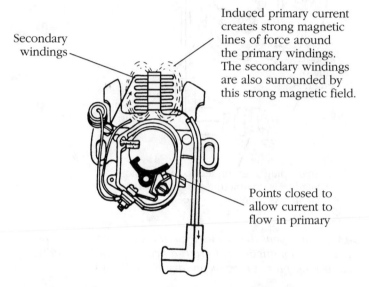

Induced primary current creates strong magnetic lines of force around the primary windings. The secondary windings are also surrounded by this strong magnetic field.

Secondary windings

Points closed to allow current to flow in primary

2-12 *Points closed: The primary coil winding is saturated with magnetic flux. Current flows through the primary windings and to ground through the stationary point.*

Figures 2-12 and 2-13 show magneto operation about as well as drawings can. As the flywheel turns, a magnet sweeps the coil to produce a voltage in the primary winding. When the breaker points close, the circuit is complete and primary current flows (FIG. 2-12). Current flow produces a strong magnetic field around the coil windings, which is independent of the field produced by the flywheel magnets.

When the points cam apart, the primary circuit loses ground and opens (FIG. 2-13). No current flows and the magnetic field

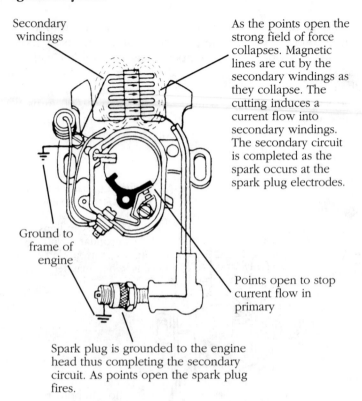

Secondary windings

As the points open the strong field of force collapses. Magnetic lines are cut by the secondary windings as they collapse. The cutting induces a current flow into secondary windings. The secondary circuit is completed as the spark occurs at the spark plug electrodes.

Ground to frame of engine

Points open to stop current flow in primary

Spark plug is grounded to the engine head thus completing the secondary circuit. As points open the spark plug fires.

2-13 *Points open: the primary circuit loses continuity. The collapse of the magnetic field around primary windings induces a high voltage pulse in the secondary. Spark plug fires.*

associated with that current collapses. The rapidly shrinking magnetic field induces a voltage in the secondary windings. Magneto open-circuit voltages peak at around 18,000 V.

The condenser acts as an electrical accumulator to provide temporary electron storage. One side of the condenser is "hot" and wired across the points; the other side grounds through the metal condenser case. The condenser charges when the points break, storing electrons that would otherwise find ground by arcing across the point gap. Microseconds later, primary voltage diminishes enough to allow the condenser to discharge to ground through the primary coil winding. The backflow of electrons from the condenser neutralizes the remaining primary voltage, speeding the collapse of the magnetic field and boosting secondary voltage.

Troubleshooting magnetos

Most magneto faults originate with the breaker points, which—sooner rather than later—fail. Normal contacts are slate gray and somewhat mottled in appearance, but without the telltale peaks and valleys associated with metal transfer. Oxidized points are dark and sometimes black. The tip of the movable arm can turn blue from overheating. Wear at the rubbing block, where the movable point contacts the cam or on the point plunger (Briggs & Stratton) will progressively narrow the point gap. After long service, the spring can weaken enough to allow the movable arm to lose contact with the cam. Ignition becomes erratic.

Some point sets can short to ground through the movable-arm spring, although this usually occurs as the result of assembly error and is readily apparent. Point contacts can also become oil-fouled as a result of crankshaft seal failure or oil seepage through the plunger used on certain Briggs & Stratton magnetos. Oil-fouling can be recognized by the splatter of carburized oil under the contacts.

Condenser failure can have any of several symptoms, including reduced spark output, misfires, increased breaker point arcing, and migration of tungsten from one point face to the other. A shorted condenser will shunt primary current to ground. Replace the condenser as an assembly with the breaker points.

Inspect primary wiring for insulation damage, which can occur if the wiring fouls against the flywheel. The coil is the last component you should suspect because it is generally reliable and always expensive. A few shops use coil testing machines, but the surest check is to substitute a known-good coil. See the "Coil and Air Gap" section.

Breaker points

Small engine point assemblies are built in two basic configurations. What we might call the "standard" configuration, found on all battery-and-coil systems and on most magnetos, consists of a pivoted point arm, a flat point spring, and a fixed arm. The movable arm can bear directly against the point cam via a nylon or phenolic rubbing block at the cam interface. Alternately, the point assembly can be mounted at some distance from the cam and articulated by means of a plunger.

Standard configuration points secure to the baseplate, or stator, with one or two screws that can be supplemented by locator pins (FIG. 2-14). When working space is tight (as, for example, when replacing points through the small window provided in Bosch flywheels) it is advisable to use a magnetic screwdriver.

Remove all traces of oil from the mounting area and lightly lubricate the cam with high-melting-point grease. Just a light smear around the full diameter of the cam is sufficient. Some cams lubricate through an oil-wetted wick, which can usually be reversed to present a fresh rubbing surface to the cam. Soak the wick in motor oil. Apply one or two drops of oil to the point pivot.

Install the point set, aligning any locating pins on the underside of the assembly with holes provided in the stator plate. Tighten the electrical connection, being careful not to twist the movable arm spring into contact with the ground. Lightly secure the holddown screw(s).

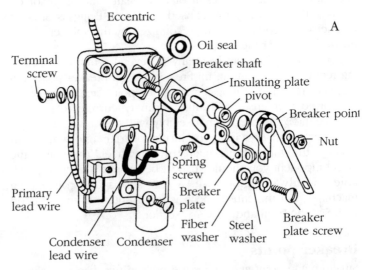

2-14 *Briggs & Stratton manufacturers these remotely mounted, standard configuration point sets as part of their Magna-Matic ignition system (A). It is a fairly complex assembly. Make careful note of the parts layout.*

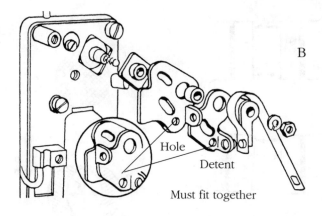

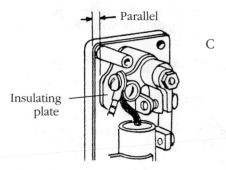

2-14 *(Continued) The detent in the stationary arm bracket must be indexed with a hole in the insulating plate (B). Upon assembly, the breaker plate must be parallel with the left-hand edge of the breaker box (C).*

While few mechanics take time to verify that contacts are parallel and concentric, this step can extend point life. Drawing A in FIG. 2-15 illustrates full contact, with both point faces meeting squarely in the same plane. Drawing B shows the effects of point misalignment. Snug down the holddown screws and correct the misalignment by adjustments to the *fixed* arm. Use long-nosed pliers or a proper bending bar, available from Tecumseh.

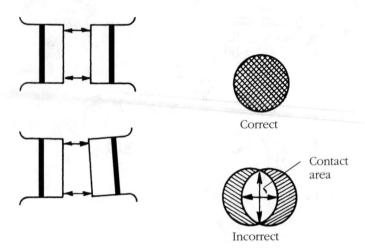

Correct

Contact area

Incorrect

2-15 *Correct point alignment results in full contact and maximum service life. Adjust by bending the fixed point arm.*

Adjust a point gap as follows:

1. Make a preliminary adjustment so the points make and break contact as the flywheel is turned. Lightly snug the hold-down screw(s).

2. Turn the flywheel until the points open to maximum extension. The rubbing block should be on the nose of the ignition cam.

3. The fastest, most accurate way to set the points is to bracket the specification with go and no-go feeler-gauge blades. In this example, we will use the common 0.020-inch specification.

   Using the adjustment screw or pry slot provided, move the stationary point until the 0.019-inch go blade slips easily between contact faces and the 0.021-inch no-go drags, forcing the movable contact open (FIG. 2-16). Feeler-gauge blades must be held dead parallel to contacts.

4. Tighten the hold-down screw(s) and check the gap, which can change. Repeat the adjustment as necessary, this time anticipating the creep.

5. Burnish contact faces with a piece of cardboard to remove fingerprints, oil, and oxidation.

6. Replace the condenser with the points, clean any oil from the condenser mounting area, and make sure the electrical lead clears the flywheel hub.

The *Briggs & Stratton point set configuration*, used on light four-cycle engines, integrates the condenser with the fixed contact. Unlike other point sets, the fixed contact is "hot" and insulated from ground by the condenser. The movable arm is grounded. Although technically obsolete, millions of these point sets remain in service (FIG. 2-17).

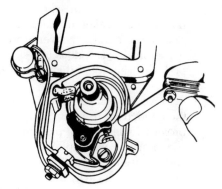

2-16 *Setting the point gap on an underfly-wheel magneto.*

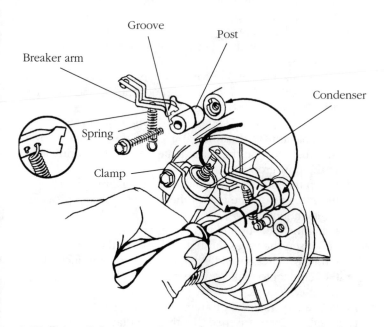

Breaker arm

Groove

Post

Condenser

Spring

Clamp

2-17 *Briggs & Stratton point configuration, used on light-frame four-cycle engines. Note the spring and ground wire orientation.*

Install as follows:

1. Remove the flywheel and point assembly cover, secured by two self-tapping screws.
2. Remove the condenser clamp screw and breaker arm screw (FIG. 2-17). Note the lay of the parts:
   ~ The braided ground strap loops under the pedestal to lie on top of the breaker arm.
   ~ The open end of the spring threads through the large hole in the breaker arm and exits through the small hole. The closed end loops over a post on the stator plate.
   ~ The breaker arm rides in a slot on the breaker-arm pedestal. The slot also indexes with a tab on the stator plate.
3. Remove the coil wire and optional kill switch wire from the condenser, using the plastic spring compressor provided with OEM replacement point sets. You can also use miniature water plump pliers to "unscrew" the spring and release the wires.
4. Oil in the point cavity means a bad crankshaft seal, or, less likely, a worn plunger bore. The bore can be reamed and bushed, if necessary.
5. Replace the plunger if it measures less than 0.870 inch long. The grooved end of the plunger goes next to the point assembly.
6. Begin assembly by indexing the tubular breaker arm post with its tab and routing the ground strap as shown in FIG. 2-17. Tighten the post screw.
7. Install the open end spring through the holes in the movable arm, as described above. Slip the closed end over the small post on the stator plate, seating the spring loop in the post groove.
8. Grasp the movable arm and, pulling against spring tension, engage the end of the arm into the slot provided on its mounting post.
9. Using the tool supplied, compress the hold-down spring and insert the wires through the hole in the condenser terminal. Wires should extend about a quarter inch out of the terminal to make square contact with the spring. Release the tool.
10. Rotate the crankshaft to retract the plunger.
11. Install the condenser so that the point faces touch and lightly snug the clamp screw.
12. Rotate the crankshaft to fully extend the plunger and open the points.

13. Adjust to the required 0.020 inch as described above.
14. Tighten the condenser clamp screw and check the gap. Some "creep" is inevitable and the operation will probably have to be repeated.
15. Burnish contacts with cardboard.
16. Install the flywheel key and flywheel.
17. Test magneto output by spinning the flywheel with the spark plug removed.

Coil and air gap

Ignition coils discussed here are used on both magneto and CDI systems. These coils receive energy from an external magnetic field and must be located in close physical proximity to that field. Battery-powered coils have similar electrical characteristics, but can be mounted anywhere a ground is available.

Modern design practice integrates the coil with its armature (or lamination pack) and, often, with other ignition circuitry. The coil can include the amplifier and pickup circuits for CDI systems and a few magneto coils include a condenser, inaccessible under the insulation.

Coil failure is evidenced by no or low spark voltage, intermittent voltage output, or (rarely) by popback through the carburetor. Heat can trigger the malfunction, although normally it is present over the whole temperature range.

Some shops use coil testers, others make resistance measurements (which vary with coil type), but the surest, least equivocal test is to substitute a known-good coil for the suspect unit.

Two or more capscrews secure the coil laminations to the stator plate. The air gap is the distance the laminations stand off from flywheel or rotor magnets.

Air gap adjustments are limited to coils that can be accessed with the flywheel in place. This category includes coil/lamination assemblies under windowed flywheels and those mounted outside of the flywheel.

The air-gap specification varies with make, model, and armature construction and should be set accordingly. Excessive air gap reduces output voltage; insufficient air gap can allow the flywheel to contact the laminations at high speed and may upset the CDI system timing enough to cause kickback during cranking.

Set the gap with the appropriate nonmagnetic feeler gauge (available from the factory or auto parts houses). Turn the crankshaft until the flywheel magnets align with the armature, insert the gauge, and loosen the armature-retainer fasteners (FIG. 2-18). The magnets will pull the armature against the gauge. Tighten the retainer bolts and remove the tool. Spin the flywheel to detect possible interference between the laminations and flywheel rim.

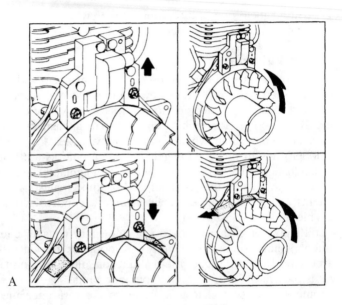

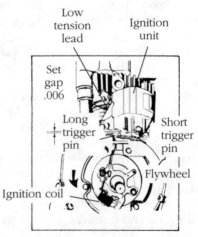

2-18 *Armature air gap adjustment for most engines is as shown in A; Repco CDI units triggers from replaceable magnetic pins that stand proud of the flywheel rim (Tecumseh application shown).*

Edge distance, E-gap, breakway gap, and pole shoe break describe the special relationship between the ignition coil armature and the magneto magnet at the moment of point break. This relationship is fixed on American engines, or if not, is clearly marked (FIG. 2-19). Figure 2-20 shows the pole shoe break for a high-performance Bosch magneto. Point break can be determined with an ohmmeter, as described under "Timing".

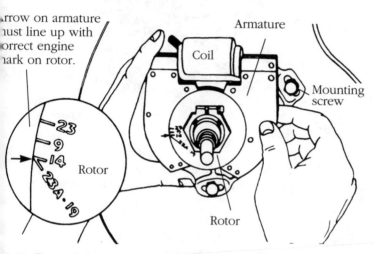

2-19 *Briggs & Stratton Magna-Matic E-gap varies with engine model and is marked accordingly.*

Optimum pole shoe break can be achieved by experimenting with point gap. The wider the gap, the narrower the pole shoe break. Thus, if the specified point gap is 0.020 inch and best spark is produced at, say, 0.026 inch, the pole shoe break should be increased until the best spark occurs at 0.020 inch.

Battery and coil systems

Still used on a number of industrial engines, battery and coil systems are the simplest to repair: parts are individually accessible (i.e., not encapsulated in black boxes) and receive power so long as the ignition switch is on. Thus the system can be tested without running the engine.

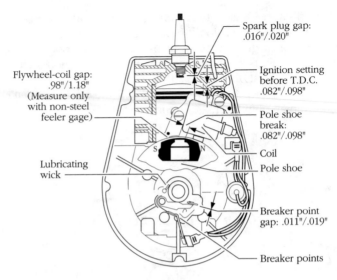

Spark plug gap:
.016"/.020"

Ignition setting
before T.D.C.
.082"/.098"

Flywheel-coil gap:
.98"/1.18"
(Measure only
with non-steel
feeler gage)

Pole shoe
break:
.082"/.098"

Coil

Pole shoe

Lubricating
wick

Breaker point
gap: .011"/.019"

Breaker points

View from PTO end

2-20 *Robert Bosch magneto has pole shoe (E-gap) specification of 0.334–0.492 inch, measured between north pole lamination and coil housing.*

Figure 2-21 is a wiring diagram of a battery and coil system that delivers voltage to two spark plugs. The primary side of the circuit consists of the battery, ignition switch, primary coil windings, breaker points, and condenser. The secondary circuit connects high voltage coil windings with the spark plugs through dual high-tension leads. Both circuits ground to the engine.

System operation is similar to magneto operation, except that battery voltage provides power for the primary circuit.

To troubleshoot this type of system, first verify battery condition, as outlined in Chapter 5. Once you are satisfied that the battery functions normally, check the point gap, which should be a tight 0.020 inch for most engines.

With the switch on and points open, determine primary circuit continuity with a test lamp, or, barring that, a screwdriver. Connect the lamp across the movable ("hot") contact arm and the fixed (grounded) arm. The lamp should come on, although dimly because some voltage goes to ground through the primary coil windings. A quick swipe of a screwdriver blade across

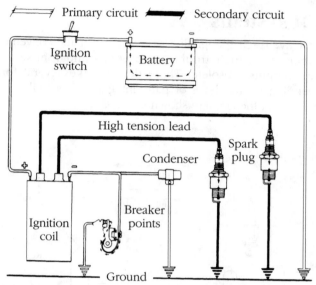

2-21 *Battery and coil circuitry for a Kohler simultaneous-firing twin-cylinder engine. If cylinders fired out of phase, a second coil or distributor assembly would be used.*

the contacts should produce sparks. If no primary voltage can be detected, use the lamp to trace circuit continuity back to the battery.

If the lamp lights when connected across the movable and grounded points, turn the flywheel until the points close. The lamp should go out because the "hot" point now grounds through the stationary contact. If the lamp remains lit, point contacts have oxidized and should be filed or replaced.

If both of these tests are negative—i.e., if primary voltage is present at the movable point arm with contacts open and absent with points closed—it can be assumed that the problem is in the secondary circuit. Usually it is, but be sure that the points function. If you have not replaced the point set, replace it now, together with a new condenser. See instructions in the "Breaker Points" section.

The secondary circuit always includes an ignition coil and can incorporate a distributor cap, rotor, and coil-to-distributor wire. Replace the less expensive parts first, reserving ignition coil replacement until last.

CDI systems

Capacitive discharge ignition systems have made conventional systems, with their troublesome points and rpm-sensitive spark outputs, obsolete. These devices take several forms, depending on the level of parts integration, but all employ variations of the circuitry shown in FIG. 2-22.

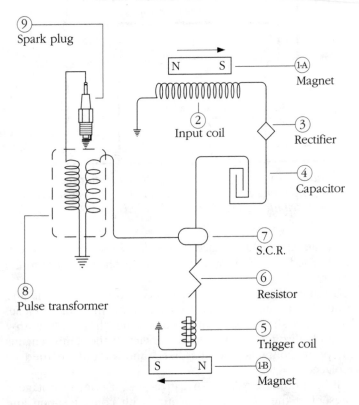

2-22 *Tecumseh CDI system is typical of the state of the art. All of these systems would benefit from a provision for automatic advance.*

The flywheel magnet (lA) generates about 200VAC in the input coil (2) at normal engine speeds. The rectifier (3) converts alternating current to direct current, which is stored in the large capacitor (4). About 180 degrees of crankshaft rotation later, the magnet passes the trigger coil (5), which generates a small signal voltage across the resistor (6). This voltage causes the silicon-controlled rectifier (7) to become conductive, releasing the charge on the capacitor (4) into the primary windings of the pulse transformer (8). A large voltage is generated in the transformer secondary winding, which goes to ground across the spark plug electrodes.

Most CDIs are covered under the same 90-day warranty as the engine. But the warranty is worth investigating, because top-of-the-line CDIs receive factory support for as long as ten years after purchase.

You must observe several cautions when you are servicing electronic ignition (and charging) systems:

• Do not introduce stray or reversed polarity voltages into the system. Reversing battery polarity is a fatal error for most systems (FIG. 2-23). Nor is it permissible to arc weld on the engine or equipment connected to the engine without first disconnecting the CDI. If you use an ohmmeter, follow factory guidelines, which limit test voltages and designate safe test points.

• Some CDI ignition/charging systems use the battery as a ballast resistor. Running the engine without the battery can create an over-voltage condition.

• Do not disturb the circuit while the engine is running or during cranking. This means that you must not disconnect or ground primary wiring. Black boxes ground through their hold-down bolts and must remain attached to the engine. Nor is it permissible to crank the engine with the spark plug lead disconnected and ungrounded. Ground the cable through a spark gap tester. Open-circuit voltages are high enough to puncture the pulse transformer.

Troubleshooting begins with a visual examination of connectors, external wiring, interlocks, and the flywheel key. Heavy rust deposits on flywheel magnets should be sanded off.

No further troubleshooting is possible when the CDI unit is integrated into a single package; you simply buy another one. More conservative designs locate the module (or ignition unit) in a relatively cool spot at some distance from the generating /trigger coils (FIG. 2-18B). Your problem is to determine which unit has failed. The module is the more likely candidate, but it is also the more expensive.

Factory-supplied resistance specifications are helpful, but not always conclusive. The best bet is to turn the problem

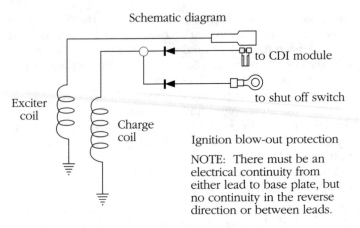

Schematic diagram

Exciter coil

Charge coil

to CDI module

to shut off switch

Ignition blow-out protection

NOTE: There must be an electrical continuity from either lead to base plate, but no continuity in the reverse direction or between leads.

2-23 *A few quality CDIs include blowout protection in the form of blocking diodes.*

over to an authorized dealer, who should have the necessary Merc-O-Tronic or Graham-Lee test apparatus—or, lacking that, parts to substitute.

Timing

Many light utility engines have no provision for timing adjustment. These engines can be recognized by fixed stator plates and the absence of timing marks.

Stator-plate rotation

A number of American-made engines use Phelon or Wico magnetos with elongated slots at the stator-plate mounting flange. These slots allow the whole assembly to be moved a few degrees relative to the piston. Two systems are used to indicate the correct stator plate position and timing. Some manufacturers provide indexing marks on the stator and engine block (FIG. 2-24). Others, such as Jacobsen, use the limit of adjustment as the timing referent. Depending upon crankshaft rotation, the stator plate is moved full clockwise or counterclockwise. Large Tecumseh engines are timed in the same way.

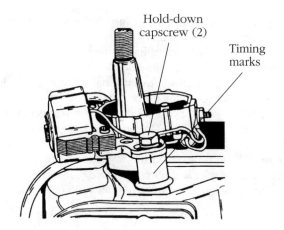

Hold-down
capscrew (2)

Timing
marks

2-24 *Wico, Phelon and other small magnetos time to by rotating the stator to align punch marks. Marks are valid for the
particular magneto/engine combination.*

Point gap variations

The wider the point gap, the sooner the points break, and the
earlier the spark. This relationship holds for all point-
equipped engines, but not all manufacturers specify it as a
part of their timing drills. Those that do have generally ar-
ranged matters so that the points are accessible with the flywheel installed. Timing marks are located on the flywheel rim
and on a stationary engine feature. Often you will encounter
two flywheel marks an inch or so apart. The first mark to align
as the flywheel is turned in its normal direction of rotation
represents top dead center. The second mark is the timing
mark.

Follow this general procedure:

- Set the breaker point gap to manufacturer's specs.
- Connect a battery-powered test lamp or ohmmeter across
  the points, one lead clipped to the movable arm and the
  other to ground. Disconnect the lead to the coil (necessary
  because the coil is grounded).
- Locate the timing marks, which can involve removing the
  shroud or an access plug (certain Kohler models). Honda
  G40 and G60 series engines use the crankcase split as the
  stationary mark and the "F" mark on the flywheel rim as
  the firing mark.

- Bring the piston up on the compression stroke so that the timing marks approximately align.
- Turn the crankshaft about half a revolution against the direction of normal rotation.
- Slowly turn the crankshaft in normal rotation until the lamp or meter indicates point break.
- Adjust point gap to synchronize point separation with timing mark alignment. The range of permissible adjustment varies with the manufacturer, but it is usually not more than plus or minus 15 percent of the nominal gap. For example, if the manufacturer calls for 0.020-inch, the actual adjustment can range between 0.017 and 0.023 inch.

Point assembly rotation

Some designs include provision for rotating the breaker points relative to the actuating cam. Wico and Phelon magnetos have this feature, because their slotted mounts allow rotation relative to the cam. But these magnetos use timing marks as the final arbiters, without reference to point break. The discussion here concerns those systems that combine point assembly rotation with point break.

Figure 2-25 shows how ignition timing is set on Onan CCK series engines. Point gap is set at 0.020 inch, a light is connected across the points, flywheel timing marks are aligned, and the point box is moved left or right to initiate point break.

Timing light

Any engine with external timing marks—that is, marks that are visible when the engine is fully assembled and running—can be timed dynamically with a strobe light. Engines with CDI ignitions or with automatic advance mechanisms must be timed with a light.

Caution: Do not operate an engine with the cooling shroud removed.

If you must buy a timing light for this drill, invest in a good one, with an easily replaceable xenon bulb and high-speed switching circuitry. Modern timing lights trigger inductively from the spark plug cable and require a source of 12Vdc negative-ground power.

Warning: Do not look directly into the strobe. Xenon bulbs are bright enough to cause retina damage.

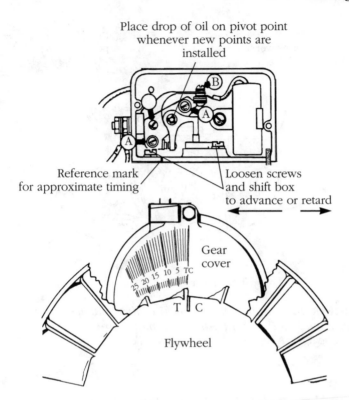

Place drop of oil on pivot point whenever new points are installed

Reference mark for approximate timing

Loosen screws and shift box to advance or retard

Gear cover

Flywheel

2-25 *Onan CCK and CCKA series engines time by moving the breaker box relative to the cam. Top dead center is indicated by the letters TC, a standard industry practice. But because Onan timing specs vary with engine model and optional equipment, no specific timing mark is provided. CCK engines fire 19 degrees before top dead center (BTDC). CCKA electric-start engines with fixed advance are set at 20 degrees BTDC; with auto advance, the specification is 24 degrees BTDC.*

Multicylinder engines time to No. 1 cylinder, usually defined as the cylinder closest to the flywheel. (Because two connecting rods share the same crankpin on V and horizontally opposed engines, one cylinder always leads.)

Engines with fixed advance are timed at idle speed (FIG. 2-26). The drill for automatic advance is more difficult to generalize. Some manufacturers (e.g., Sachs) provide two timing marks, one for idle rpm and the other for full advance. A few, such as Onan, provide a full-advance mark only, which means that timing must be accomplished under full advance (FIG. 2-27).

Timing sight hole
(or bearing plate
or blower housing)

Changing
point gap

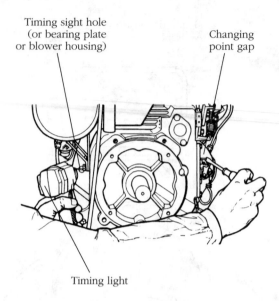

Timing light

2-26 *Kobler single-cylinder timing drill involves small adjustments to the point gap while marks are "frozen" with a strobe light. Normally this is done at idle.*

Point break/piston position

Any magneto-equipped engine for which specifications are available can be timed with reference to piston position BTDC, although this procedure should not be considered part of normal maintenance. Rarely are industrial and utility engines timed with this exactitude.

You need a dial indicator, rigged to work through the spark plug port (FIG. 2-28). Set the point gap to specification and connect a continuity reading device across the points, as described under "Point gap variations."

Locate top dead center, nudging the flywheel in ever smaller increments until the piston is at its point of zero motion. Once this is done, zero the dial indicator. Turn the flywheel in the normal direction of rotation until the points break. Note the indicator reading, which should be within one percent or so of the timing specification. Correct by rotating the point assembly relative to the crankshaft. If that is not possible, vary the point gap. Mark the flywheel and engine block so that the engine can be timed dynamically in the future.

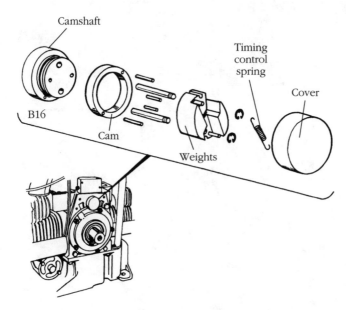

2-27 *Onan and other automatic advance mechanism occasionally need cleaning or spring replacement. Engines so equipped must be timed while running at relatively high speed (1500 rpm for Onan).*

2-28 *Tecumseh supplies a dial indicator setup for small engine work as PN 670241.*

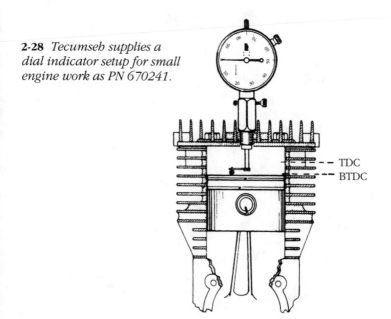

3

Fuel system

The fuel system consists of the carburetor, air cleaner assembly, fuel tank and miscellaneous fittings (including the tank strainer), and shutoff valve. A fuel pump might also be included in the package.

How carburetors work & fail

We think of an engine as a source of power; it is equally true to say that it is a vacuum pump. The partial vacuum developed during the intake stroke sets up a pressure differential across the carburetor. Air and fuel, impelled by atmospheric pressure, move through the instrument to equalize pressures.

Every carburetor has a *venturi*, or a necked-down section of the bore upstream of the throttle plate. When the airstream encounters this restriction, it speeds up and simultaneously loses pressure. The venturi creates a local low pressure zone.

Fuel, under atmospheric pressure, moves from the carburetor reservoir through the main jet and through the nozzle, which opens into the low pressure area created by the venturi. The jet can be fixed, as shown in FIG. 3-1A, or it can be adjustable. In any event, jet orifice area determines how much fuel is mixed with the air in the carburetor bore. The main jet, interconnecting passages, and fuel nozzle make up the high-speed circuit—"high speed" because this circuit flows only when the throttle blade is open. Closing the throttle blocks the air supply to the venturi and the circuit shuts down.

A closed or nearly closed throttle represents a major restriction—or, if you will, a kind of crude, unstreamlined venturi. Very low pressures develop near the trailing edge of the blade. Low-speed circuit ports discharge fuel into this low pressure area. The port nearest to the engine, called the primary idle port in FIG. 3-1B, functions when the throttle is against its stop; secondary idle ports progressively come on-stream as the throttle opens (FIG. 3-1C). The transition between low- and high-speed circuits occurs at about one-quarter throttle.

Like many modern designs, the carburetor shown has fixed jetting, with no provision for adjustment. Air bleeds emulsify the fuel before delivery and help prevent siphoning.

Older and all foreign carburetors include a low-speed adjustment screw, normally located near the throttle blade. This screw, when found adjacent to the throttle blade, controls fuel delivery. Turning it out richens the mixture at low throttle angles.

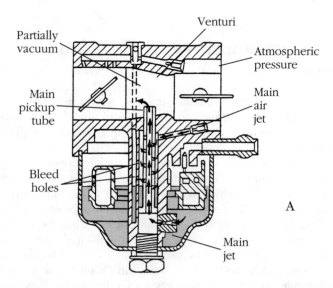

3-1 *During high-speed operation, fuel discharges into the venturi through the main pickup tube, sometimes called the main nozzle (A). An air bleed emulsifies the fuel, breaking it into droplets, prior to discharge and atomization.*

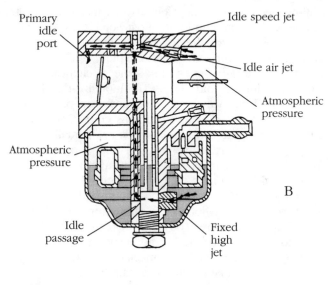

Primary idle port

Idle speed jet

Idle air jet

Atmospheric pressure

Atmospheric pressure

Idle passage

Fixed high jet

B

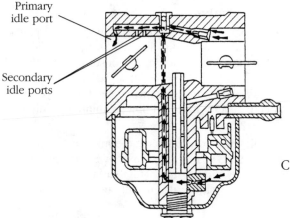

Primary idle port

Secondary idle ports

C

3-1 *(Continued.) The low-speed circuit discharges into the primary idle port (B) and, at larger throttle angles, into secondary, or off-idle, ports (C).*

Note: A few obsolete American and some European carburetors control low-speed mixture by varying the amount of air delivered. The adjustment screw can be recognized by its blunt, rounded end (fuel adjustment screws are sharply pointed) and can be located anywhere on the instrument, even under the air cleaner.

A high-speed adjustment might also be included. The high-speed screw, mounted under the float bowl or on the carburetor body upstream of the low-speed screw, regulates fuel delivery, as shown throughout this chapter.

Virtually all carburetors employ some mechanism to richen the mixture during cold starts. Traditionally this has been done with a choke valve, mounted upstream of the venturi. When the choke is closed, all circuits come under vacuum and flow. The choke can be automatic or manually operated. A newer approach is to use a small pump, or primer. Wet primers deliver a squirt of raw fuel into the main nozzle; dry primers use air pressure to temporarily flood the carburetor.

The internal fuel level must be maintained within narrow limits. Too much fuel floods the carburetor and, in the process, richens the mixture. Too little fuel results in a lean condition and "dead spots" during acceleration. Carburetors are classed by their fuel-regulating mechanisms, which can take the form of a float, diaphragm, or pickup tube.

Float & inlet valve

The float mechanism consists of a hollow metal or plastic float and an inlet valve assembly, known as the needle and seat (FIG. 3-2). As fuel in the reservoir is consumed, the float drops, allowing the needle to fall away from its seat. Fuel enters the chamber until the float rises and closes the valve, an action that can occur several hundred times a minute at full throttle.

Flooding results if:

- The needle fails to seal against the inlet seat. This condition can be caused by simple wear or, more commonly, by dirt on the mating surfaces.
- The inlet-seat gasket leaks, usually because the inlet seat has vibrated loose.
- The float sticks open. This condition is most likely to occur upon refueling a dry tank. Sometimes it can be cured by loosening the bowl nut and rotating the float cover a fraction of an inch in either direction.
- The float sinks. Metallic floats can develop leaks, and plastic foam floats can become fuel-saturated. In either case, replace the float.

Fuel will be denied if:

- The needle sticks, which can be caused by varnish accumulations on the needle or by improper assembly of the float spring.

- The float is set too high. See the "Repair" section for float height adjustments.
- The float bowl vent is clogged. This condition is more likely to occur in carburetors with dry primers that, when activated, pressurize the float chamber.

Diaphragm

Carburetors intended to be operated off the horizontal employ a diaphragm-controlled inlet valve (FIG. 3-3). The lower side of

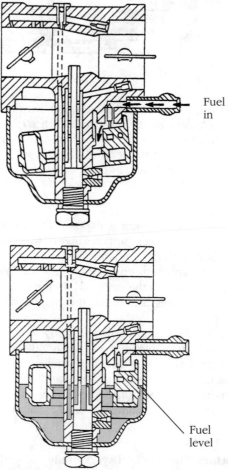

Fuel in

Fuel level

3-2 *The float mechanism maintains a preset fuel level in the float chamber and above the main jet. When failure occurs, it is generally associated with the inlet valve.*

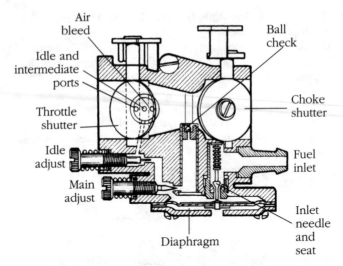

3-3 *A Tecumseh-pattern diaphragm carburetor. Cover vent and diaphragm rivet (referenced in text) are clearly shown. Mixture adjustment screws are critical. Overtightening the idle adjust screw can twist its tip off into the jet, where it is almost impossible to extract. Screws have the same thread, but interchanging them does irreparable damage to the carburetor.*

the neoprene diaphragm is exposed to atmospheric pressure; the upper side "sees" intake manifold pressure, which is always lower than atmospheric. The pressure differential moves the diaphragm upward and lifts the inlet needle off its seat. Fuel flows into the cavity above the diaphragm. As fuel is depleted, the pressure above the diaphragm drops and the cycle repeats.

Some of these carburetors employ a check valve in the high-speed circuit (as shown), apparently to prevent siphoning. Others have an air-operated primer in lieu of the choke. Depressing the primer bulb raises the diaphragm, flooding the instrument. Choke-equipped versions can be deliberately flooded by gently pushing up on the diaphragm rivet with a matchstick inserted through the vent port in the diaphragm-chamber cover.

Caution: This method works only for carburetors with a vent hole in the center of the cover.

Bendix and certain Tillotson models have a second diaphragm, used as a fuel pump. The pump diaphragm can be recognized by the two flaps cut into it, which serve as check valves.

Diaphragm carburetors generally fail during starting and especially after wet storage. Long exposure to gasoline hardens the diaphragm and leaves varnish deposits that stick the needle in the closed position. Replace these parts.

Suction lift

Practically speaking, these suction lift carburetors are a Briggs & Stratton monopoly (FIG. 3-4), although all American manufacturers have experimented with them. Newer Briggs models, known as Pulsa-Jets, include a diaphragm-operated fuel pump that discharges into a cup built into the tank. If the cup is dry, several pulls of the starter cord are required to start the engine. Replace the diaphragm whenever the carburetor is serviced. Pickup tube screens and check valves fail if the fuel is allowed to dry in the tank. A single mixture-adjustment screw on the side of the instrument regulates both high- and low-speed mixture compositions.

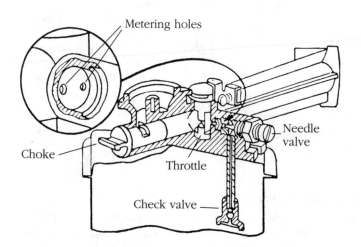

3-4 *Briggs & Stratton suction-lift carburetor. The newer version, shown in FIG. 3-21, features an integral fuel pump and revised choke.*

Troubleshooting

Do not rush into carburetor work. Test ignition system output with a spark-gap tool and install a known-good spark plug. Verify that the fuel is fresh and uncontaminated by water or rust particles. Mix fuel with the correct proportion of oil for two-cycle engines.

Check the air filter and clean or renew the element as described in the "Filter" section.

Warning: The air filter doubles as a flame arrestor. Do not attempt to start or run an engine with the filter element removed.

And, finally, make certain that the choke (when fitted) closes and opens fully (FIG 3-5). Most engines will not start unless the choke effectively seals off the carburetor bore; none will run properly if the choke remains partly closed.

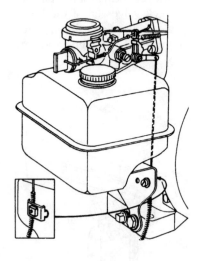

3-5 *A misadjusted cable clamp can deny full choke on many applications. Briggs & Stratton configuration is typical.*

No fuel delivery

Zero fuel delivery is obvious because the spark plug tip remains resolutely dry after prolonged cranking. Heroic efforts might oil the tip on four-cycle engines, but the characteristic odor of gasoline is absent. The carburetor bore, visible when the air cleaner is removed, is dry or lightly wetted. A definitive test is to spray carburetor cleaner into the spark plug port. If the engine starts but refuses to continue running, the problem is fuel starvation.

Pickup-tube and diaphragm failures are characteristic of carburetors that use these components. Float-type carburetors rarely fail to pass fuel, unless disabled by exposure to water or stale gasoline. The problem is more often upstream of the float carburetor: Check the tank screen and, when applicable, filter and fuel pump. Cracking the fuel line at the carb connection should yield a dribble of gasoline from gravity-fed systems; pump-fed systems must, of course, be activated by cranking the engine.

Warning: Fuel spills are dangerous. Perform this test in a well-ventilated area, preferably outdoors, with ignition OFF.

Carburetor flooding

Any carburetor will flood and dribble raw gasoline from the air cleaner area if overprimed or overchoked. But self-induced flooding is a serious and extremely hazardous malfunction.

I have never seen a suction-lift carburetor flood spontaneously; a diaphragm carburetor can flood, but the condition is extremely rare; float-type carburetors flood regularly. The problem can normally be traced to the float or inlet valve.

Refusal to idle

Refusal to idle can be caused by restricted throttle plate movement or by an obstruction in the carburetor low-speed circuit. Failure of the throttle plate to close can be caused by:

- Idle rpm set too high. Adjust the throttle stop screw as necessary (FIG. 3-6).
- Misadjusted throttle cable (illustrated in FIG. 3-5): Loosen the clamp screw and move the Bowden cable as necessary to restore idle function.
- Binding throttle linkage or throttle-blade shaft.
- Malfunctioning or maladjusted governor (see "Governor" section).

If the engine dies when the throttle is closed, the problem is almost certainly an obstruction in the low-speed circuit. As described in "Refusal to run at high speed," most carburetors are fitted with an adjustable low-speed jet. Back out the adjustment screw to increase fuel flow. If the problem cannot

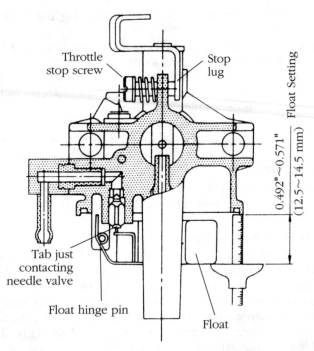

3-6 *All carburetors regulate idle rpm with a throttle stop screw, usually located on top of the instrument. Drawing also includes the float height specification, which would be determined with the carburetor inverted.*

be adjusted out, dismantle and clean the instrument. Note that improper assembly of Walbro carburetors can result in loss of idle.

Refusal to run at high speed

Failure to run at large throttle angles can have several causes, including misadjusted throttle cable, binding throttle linkage or throttle shaft, and misadjusted or failed governor. Weak throttle springs are high on the list of possibilities.

Warning: Replace failed throttle springs with the correct PN part. Otherwise, the engine may overspeed.

Vacuum leaks, downstream of the throttle plate, produce the same symptom. Check the carburetor mounting flange bolts and gasket integrity. Two-cycle engines can develop air leaks at the crankcase seals, which deny full throttle unless the choke valve is closed.

Adjustable main jets cover a multitude of sins—including partially blocked high-speed circuits, restrictive air cleaners, and vacuum leaks—but large adjustments should not be necessary. A healthy carburetor holds adjustment over the life of the engine with no more than an occasional tweak.

Black smoke, wet exhaust

These symptoms point to an overly rich mixture. The problem can be caused by a restricted air cleaner, partially closed choke, or misadjusted jets. Fixed-jet engines can smoke at high altitudes. See the "Adjustment" section.

Although unlikely, gasoline-rich mixtures can result from failure of the float or diaphragm to regulate the internal fuel level. Normally in these cases, the carburetor floods and leaks gasoline from the air horn, resulting in power loss and surging.

Stumble under acceleration

This catchall category includes symptoms associated with a dirty or leaned-out carburetor. Clean the carburetor and make the adjustments as described later.

Removal & installation

The carburetor bolts to the cylinder head or directly over the fuel tank. The fuel supply must be shut off on gravity-feed systems.

Warning: Carburetor work inevitably involves some gasoline spillage. Work in a well-ventilated place, preferably outdoors, and at a safe distance from potential ignition sources.

The governor mechanism must be disengaged from the throttle arm without doing violence to associated springs and wire links. Most springs have open-looped ends and can be disengaged with needle-nosed pliers. Closed-loop springs can be coaxed out of their mounting holes with a gentle twist. When multiple holes are provided, note—and, if necessary, mark—the hole used. Leave the wire links in place for now.

Remove the air cleaner and unbolt the carburetor at its mounting flange. Holding the carburetor in one hand, twist and rotate it out of engagement with the governor links, being careful not to bend the links in the process. Clean the gasket surfaces.

Warning: Some flange gaskets appear to contain asbestos. Remove gasket fragments with a single-edged razor blade and dispose of the material safely. Do not use a wire wheel or any method that would generate dust.

Warped suction-lift tank flanges (FIG. 3-7) can be repaired with Briggs & Stratton PN 391413. Cracked or leaking tanks cannot be repaired and must be replaced.

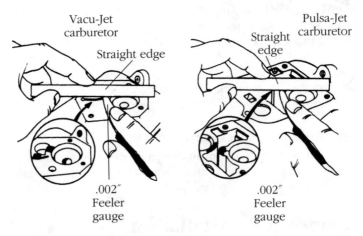

Vacu-Jet carburetor

Straight edge

Pulsa-Jet carburetor

Straight edge

.002″ Feeler gauge

.002″ Feeler gauge

3-7 *Use a straightedge and feeler gauge to check tank flange trueness (suction-lift carburetors). Discard the tank if cracked.*

Installation is the reverse of removal. Connect the governor links first, make up the flange bolts over a new gasket, and conclude with the springs. Briggs & Stratton suction-lift carburetors with automatic chokes introduce some complexity into the process (FIG. 3-8).

Repair

Carburetors do not wear out in the accepted sense of the term and should not require a massive transfusion of parts to be made operable. Rarely do you need to replace more than the needle and inlet seat, diaphragm, cover gasket and, possibly, mixture-adjustment screws. These parts can be installed without major disassembly.

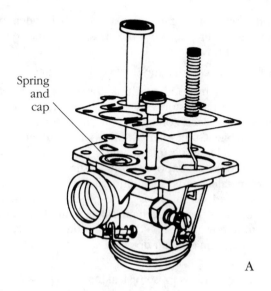

Spring
and
cap

A

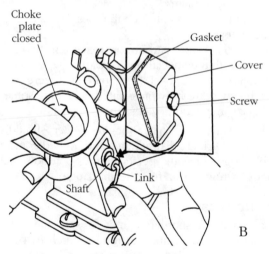

Choke
plate
closed

Gasket

Cover

Screw

Link

Shaft

B

3-8 *Briggs & Stratton automatic choke diaphragms are some-*
what tricky to assemble. Invert the carburetor body and install
the spring and cap, followed by the choke diaphragm (A). Still
holding the carburetor upside down, lower the tank into
position. Turn the assembly over and thread the mounting
screws in about two turns, or just enough to hold. Connect the
choke-actuating link. Tighten the screws, drawing them down
evenly a few turns at a time, while the choke blade held shut
(B). The choke should open when released.

Slightly dirty carburetors can be cleaned with lacquer thinner and compressed air. Blow out the passages in the reverse direction of normal flow.

Caution: Do not pressurize the float or diaphragm(s).

More severe cases respond to Gunk Carburetor Cleaner—available from auto parts stores in pint containers—and very corroded carburetors need something on the order of Bendix Econo-Cleane. While Gunk seems fairly benign, Bendix and other commercial cleaners rapidly attack plastic and rubber parts and, with prolonged exposure, zinc castings.

Strip off nonmetallic parts, including gaskets, O-rings, elastomer inlet seats, and diaphragms. Some plastic parts, such as the nylon check valves used on certain Tillotson models, require major surgery to remove and should therefore be left in place. A few minutes in the chemical bath does not seem to hurt.

Do not remove:

- Expansion plugs (unless loose or leaking)
- Throttle and choke plates
- Briggs & Stratton suction-lift pickup tubes (except for replacement)
- Main nozzles on Walbro carburetors (except for replacement with the special service part)
- Any part that resists removal. Many parts that were formerly threaded are now press-fitted, including some fuel inlet fittings, some main nozzles, and some jets.

A rebuild kit, available from the engine dealer, should contain new gaskets, O-rings, diaphragms, and a needle and inlet seat assembly, together with specifications. Some kits include mixture control screws and a float height gauge.

Float carburetors

Figure 3-9 can serve as a shorthand guide for servicing float-type carburetors, although there are structural differences between makes and models. The sidedraft carburetor shown mounts horizontally, with the float housed in a bowl under the instrument. A central nut, which might be combined with the main jet mixture control screw, secures the float bowl to the casting. Fuel moves up from the chamber to the carburetor bore—or, as it is sometimes called, the air horn.

Downdraft and updraft carburetors mount vertically, which affects the parts arrangement. Zenith downdraft carburetors (FIG. 3-10) follow automotive practice and locate the float

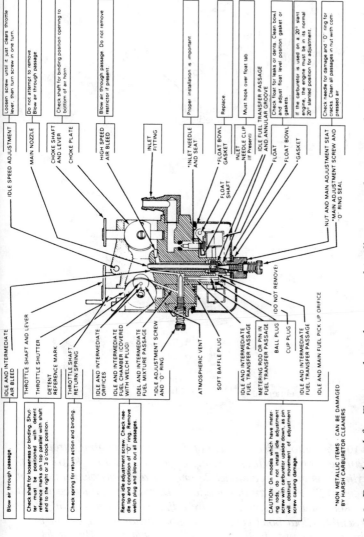

IDLE AND INTERMEDIATE AIR BLEED — Blow air through passage

THROTTLE SHAFT AND LEVER — Check shaft for looseness or binding. Shutter must be positioned with detent reference marks on top parallel with shaft and to the right or 3 o'clock position.

THROTTLE SHUTTER

DETENT REFERENCE MARK

THROTTLE SHAFT RETURN SPRING — Check spring for return action and binding

IDLE AND INTERMEDIATE ORIFICES

IDLE AND INTERMEDIATE FUEL CHAMBER (COVERED WITH WELCH PLUG) — Remove idle adjustment screw. Check needle tip and condition of "O" ring. Remove welch plug and blow out all passages.

IDLE AND INTERMEDIATE FUEL MIXTURE PASSAGE

*IDLE ADJUSTMENT SCREW AND "O" RING

ATMOSPHERIC VENT

SOFT BAFFLE PLUG

IDLE AND INTERMEDIATE FUEL TRANSFER PASSAGE

METERING ROD OR PIN IN FUEL TRANSFER PASSAGE (DO NOT REMOVE)

BALL PLUG

CUP PLUG

IDLE AND INTERMEDIATE FUEL TRANSFER PASSAGE

IDLE AND MAIN FUEL PICK UP ORIFICE

CAUTION: On models which have meter ring rods, do not install idle adjustment screw with carburetor upside down, as pin will obstruct movement of adjustment screw causing damage.

IDLE SPEED ADJUSTMENT — Loosen screw until it just clears throttle lever; then turn screw in one turn.

MAIN NOZZLE — Do not attempt to remove Blow air through passage

CHOKE SHAFT AND LEVER — Check shaft for binding position opening to bottom of air horn.

CHOKE PLATE

HIGH SPEED AIR BLEED — Blow air through passage. Do not remove restrictor if present.

INLET FITTING

INLET NEEDLE AND SEAT — Proper installation is important

*FLOAT BOWL GASKET — Replace

FLOAT SHAFT

INLET NEEDLE CLIP (If Present) — Must hook over float tab

IDLE FUEL TRANSFER PASSAGE AND ANNULAR GROOVE

FLOAT — Check float for leaks or dents. Clean bowl and adjust float level position gasket or gaskets.

FLOAT BOWL — If the carburetor is used on a 20° slant engine, the engine must be in its normal 20° slanted position for adjustment.

*GASKET

NUT AND MAIN ADJUSTMENT SEAT

*MAIN ADJUSTMENT SCREW AND O' RING SEAL — Check needle for damage and O' ring for cracks. Clean all passages in nut with compressed air

*NON METALLIC ITEMS CAN BE DAMAGED BY HARSH CARBURETOR CLEANERS

3-9 *Developed for Tecumseh carburetors, this illustration has general application to other float-type units.*

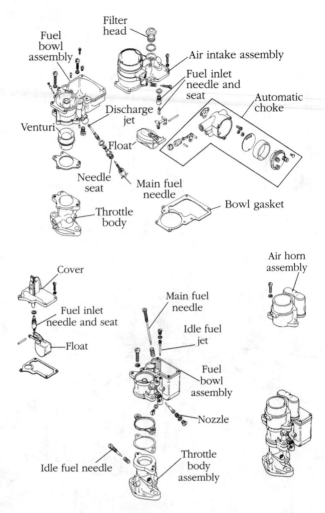

3-10 *The automotive heritage of Zenith small engine carburetors can be seen from these examples fitted to Kohler K662 twin-cylinder engines. The Model 28 employs a removable venturi, an integral filter, and an automatic choke that could have been built by Rochester. Low-speed mixture adjustment is on the air intake assembly (float bowl cover). The more usual arrangement is for this adjustment to be adjacent to the throttle plate. The Model 228 has the idle mixture control, or idle fuel needle, in the customary location. The throttle body assembly, common to both models, is not intended to be dismantled for repair. When the throttle shaft is worn, the assembly must be replaced as a unit.*

chamber alongside the air horn. Onan downdraft and Briggs & Stratton updraft designs route the air horn through the float chamber (FIGS. 3-11 and 12). The former use double pontoon floats; the latter, conventional doughnut floats.

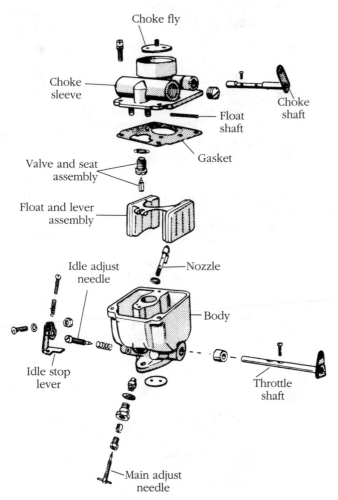

3-11 *Downdraft carburetor used on Onan CCK/CCKA twin-cylinder engines. The main nozzle consists of a brass tube, athwart of the air stream. Air enters through radial holes in the nozzle to emulsify fuel before delivery. Replaceable throttle shaft bushings indicate that the carburetor is intended for long service. Double-pontoon floats must be level and parallel.*

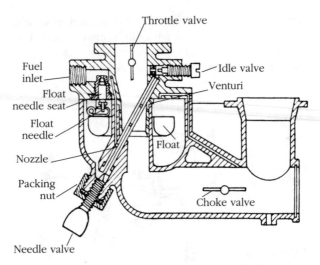

3-12 *Flo-Jet downdraft carburetor in cutaway view. In Briggs & Stratton nomenclature, this is a "two-piece" carburetor because it consists of two castings, the upper, or throttle body, and the lower, or carburetor body.*

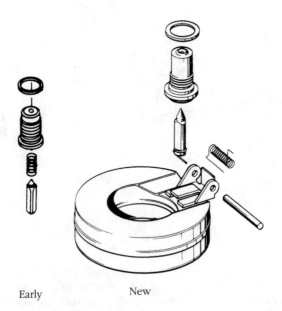

Early   New

3-13 *Two Walbro Viton-tipped needle and brass seat assemblies. Note the float dampening spring shown on the right and discussed in caption with FIG. 3-15.*

Needle and inlet seat   A few carburetors still use metallic needle and inlet seat assemblies, but the trend now is to substitute plastic for one or both parts. Figure 3-13 shows early and current Walbro elastomer-tipped needles with brass seats. Forcing the needle into the seat, as when making float adjustments, can deform the tip and result in a fuel leak. Figure 3-14 shows elastomer seats used by two U.S. makers.

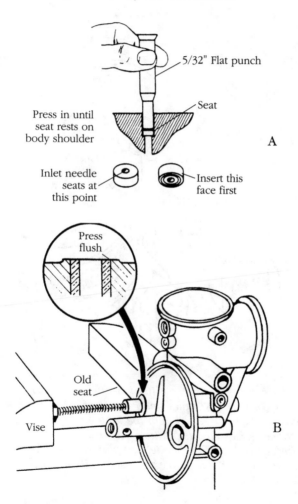

3-14 *Tecumseh Viton seats are removed with a hooked wire and installed to depth with a flat punch (A). Briggs & Stratton seats extract with a self-tapping screw and press in flush, using the original as a cushion (B).*

Tecumseh carburetors use, in various combinations, float dampener springs, needle buffer springs, and needle spring clips. The buffer spring slips over the needle, either end up. The clip and dampening spring install as shown in FIG. 3-15; the buffer spring, shown on the left of FIG. 3-13, slips over the needle, either end up.

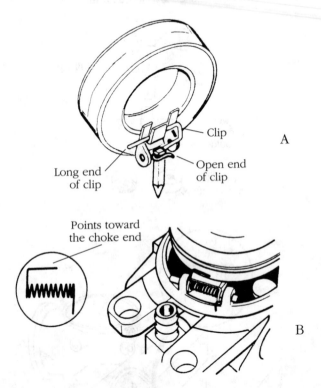

3-15 *Springs and spring clips usually follow an assembly protocol. Drawing A shows one version of an inlet needle clip, which should be installed with the long end of the clip toward the choke. Reversed installation can hang the float. Certain Tecumseh and Walbro models employ a dampening spring, positioned as shown in B to exert a slight lift on the float.*

Float height   The position of the float when the inlet valve closes must be set whenever the needle and seat have been changed. Mechanics formerly set the float level with the casting. Although this rule still holds for the Tillotson Model E and vintage numbers such as Carter Model N, it no longer has much currency.

Float height is controlled by bending the tab labeled A in FIG. 3-16. If the adjustment is minor, it can be made with the float assembled, using a small screwdriver inserted into the bend of the tab. Larger adjustments require disassembly.

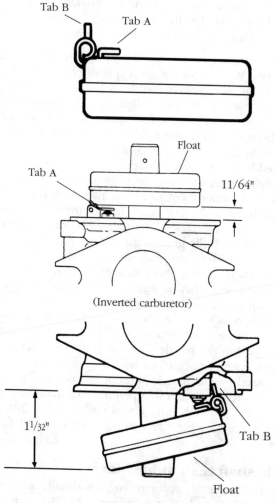

3-16 *Tab A controls float height; tab B sets float drop, but is omitted on some designs. The float height specification does not always refer to the gap between the inverted float and the (gasketless) casting. It can include the width of the float itself, as shown back in FIG. 3-6.*

Caution: Do not force the needle into contact with the seat. Elastomer parts can take a permanent set.

Float drop is the distance the float falls when the carburetor is in its normal attitude. Some designs determine this by geometry or by the torsion exerted by a dampening spring. Others use the tab shown.

Sidedraft float bowls can have an indentation that orients with the float hinge. Lay the bowl gasket flat into the lip in the underside of the main casting, stretching it as necessary.

Main nozzle   Removable nozzles have screwdriver slots or wrench flats. Walbro nozzles are a special case in that they are removable but not easily reusable. A tiny hole, visible about halfway up the threads and drilled after the nozzle was first installed, bleeds fuel into the low-speed circuit. Once the nozzle has been disturbed, the hole no longer indexes. Standard practice is to use the service nozzle, shown in FIG. 3-17 and recognized by the annular groove.

You might prefer to reuse the original part, which you can do if the main adjustment needle is not scored from forced contact with the main jet orifice. Extract the small brass cup located at nozzle thread depth on the fuel pickup pedestal. The tang end of a small file ground flat makes an appropriate tool. The lowermost of the two plugs shown in FIG. 3-17 is the one under discussion. Set the plug carefully aside, and screw in the original nozzle to within an eighth of a turn or so of seating. Gently insert a fine wire (a wire brush is a good source) into the cup plug boss while slowly turning the nozzle in and out. You will be able to sense when the wire enters the port. At this point, the port in the nozzle aligns with the low-speed circuit. Withdraw the wire and carefully tap the plug home.

I have used this technique without problems. The original nozzle seems to give better idle performance than the replacement, which is probably why it was drilled in the first place. However, I must emphasize that this repair is not factory-authorized.

Throttle shaft & shaft bushings

Carburetors fitted to heavy-duty engines usually support the throttle shaft on brass bushings, shown in FIG. 3-11, which you should replace when side play exceeds 0.006 inch. Mark the choke side of the throttle plate to ensure correct assembly. Remove the old bushings with an appropriately sized tap or EZ Out—usually ¼ inch—and press in the replacements. Ideal-

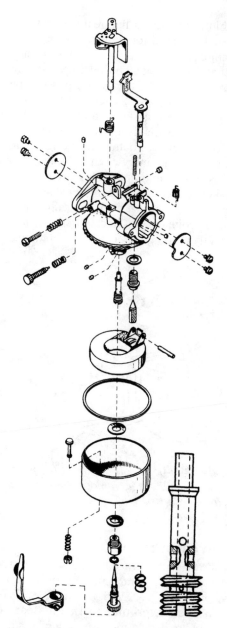

3-17 *Walbro LMG carburetor, as supplied to Clinton. Similar LMB and LMV models are used on other American engines. Groove in the replacement nozzle, shown larger than scale on the lower right, makes concern about idle-port alignment superfluous.*

ly, the bushings should be line-reamed, but this step can be omitted. Install a new throttle shaft and lightly run down the blade-mounting screws. Shut the throttle, centering the plate in the carburetor bore. Tighten the screws, staking them if that is factory practice. Test the throttle for possible binds.

Mounting flanges are usually secured to the cylinder by two capscrews. Because the gasket is thick and resilient, overtightening the screws invariably distorts the casting. The gasket surface can be restored with an "Armstrong surface mill." Tape a sheet of medium-grit emery cloth to a piece of plate glass or to a drillpress work table and, using a circular motion, grind until the gasket surface takes a uniform shape.

Float chamber covers used on vertically mounted carburetors warp if overtightened. Check as shown in FIG. 3-18 and correct by gently tapping the "ears" back into alignment.

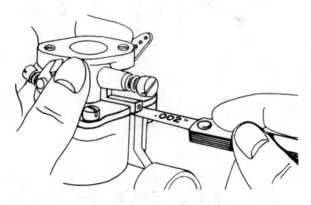

3-18 *Install the float-bowl cover without the gasket. If a 0.002-inch feeler gauge will enter, remove the cover and straighten with light hammer taps.*

Diaphragm carburetors

Figure 3-19 shows repair procedures for the most widely used diaphragm carburetor. The same carburetor is shown in exploded view in FIG. 3-20. Remove the inlet seat with a six-point ⁹⁄₃₂-inch socket, ground to fit the seat counterbore (screwdriver slots milled in the seat generally strip out before the seat budges). Replace the inlet needle and seat as a matched assembly.

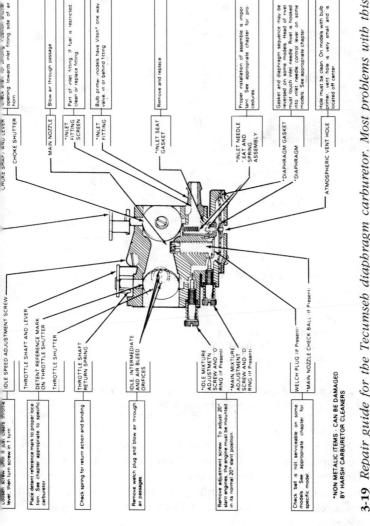

3-19 Repair guide for the Tecumseh diaphragm carburetor. Most problems with this design would be eliminated if owners ran the carburetor dry before storage.

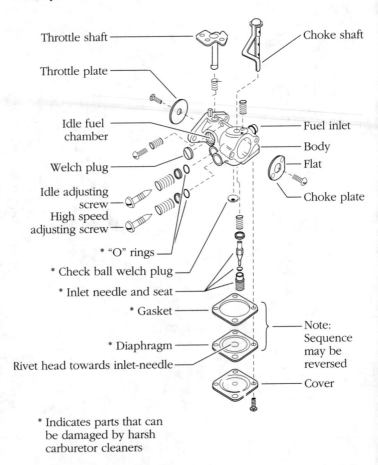

Throttle shaft

Choke shaft

Throttle plate

Idle fuel chamber

Fuel inlet

Body

Welch plug

Flat

Idle adjusting screw

Choke plate

High speed adjusting screw

* "O" rings

* Check ball welch plug

* Inlet needle and seat

* Gasket

Note: Sequence may be reversed

* Diaphragm

Rivet head towards inlet-needle

Cover

* Indicates parts that can be damaged by harsh carburetor cleaners

3-20 *Tecumseh carburetor in exploded view. The cover gasket/ diaphragm relationship is reversed on certain models.*

Most versions assemble with the diaphragm gasket above the diaphragm; those with an "F" stamped near the air cleaner flange go together with the diaphragm against the main casting, followed by the diaphragm gasket and cover. In all instances, the diaphragm rivet head is up, the splayed side down.

Inlet fuel fittings are press-fitted into the carburetor body and not disturbed unless the screen, sometimes included in the assembly, is clogged. Twist and pull the fitting out. Press in the replacement to half depth and coat the exposed portion of the shank with Loctite A. Press to depth, bringing the shoulder into contact with the carburetor body.

Caution: Low- and high-speed mixture screws do not interchange.

Suction lift

Briggs & Stratton Pulsa-Jet carburetors are distinguished from Clinton, Craftsman (Tecumseh) and earlier B&S suction-lift types by an internal fuel pump (FIG. 3-21).

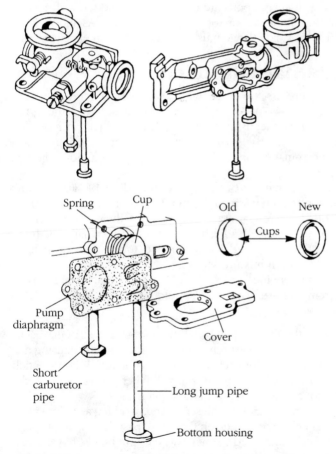

3-21 *The Pulsa-Jet, a suction-lift carburetor with integral fuel pump, is manufactured in two styles. Simple design makes it the most reliable of small engine carburetors and the easiest to repair.*

At the first sign of trouble, replace Briggs & Stratton fuel pump diaphragms and verify that pickup tube check valves work by alternately blowing and sucking on the pickup tube(s). The check ball usually sticks in the closed position and can be freed by inserting a wire up through the screened end of the tube. Plastic and brass pickup tubes press into the carburetor body.

Note: Ancient Briggs & Stratton suction-lift carburetors feed from a threaded brass tube.

Remove the mixture control screw. Using a flashlight, verify that the discharge ports are open. Replace the screw if ringed or bent. Install the carburetor using new gaskets. As shown in FIG. 3-7, the tank-side flange must be flat to make an effective seal.

External adjustments

All carburetors are fitted with an idle stop screw and most include a low-speed mixture-adjustment screw. Although the modern tendency is to use fixed high-speed jets, in most cases smaller or larger jets can be substituted for the original. Older American-made and all foreign carburetors employ adjustable main jets.

I assume that the carburetor is fully adjustable and that mixture control screws regulate fuel (and not air) delivery. Backing off the screws retracts them out of their jets and richens the mixture. Initial adjustment varies, but a ballpark figure, adequate for starting, is one or two turns open from lightly seated for both mixture screws.

Run the engine for 15 minutes or so to reach stable operating temperature. The tank should be about half-full. Follow this procedure:

1. With the throttle plate slightly more than half open, back off the high-speed screw in small increments, pausing a few seconds for each adjustment to be felt. When the rich limit is reached, engine rpm will falter and puffs of black smoke might issue from the exhaust.
2. Note the position of the screw and, working as before in small increments, tighten the screw past the region of smooth running to the point where rpm builds, hovers, and drops. This is lean roll, representing the least amount of fuel that will support combustion. Note the position of the screw.
3. Open the screw to the midpoint between lean roll and rich limit.

4. Close the throttle and adjust the low-speed mixture control for fastest idle. It might be necessary to change the adjustment of the idle stop screw.
5. Low- and high-speed adjustments are to some degree interdependent. Open the throttle as before and adjust the high-speed screw for best rpm. Drop the throttle back to idle and recheck the low speed adjustment.
6. With the engine idling, blip the throttle open.

Caution: Utility engines—particularly those with splash lubrication—require a minimum 1600 rpm idle speed to assure oil circulation. Some two-cycle engines idle at 2000 rpm. Set with a tachometer.

7. Should the engine hesitate, open the high-speed needle an eighth of a turn and retest until acceleration is rapid and smooth.
8. The high-speed mixture can be fine-tuned by making final adjustments under load.

Suction-lift carburetors use a single screw for both high- and low-speed mixture control. Models without integral fuel pumps are sensitive to the level of fuel in the tank, which should reflect normal operating conditions.

Warm up the engine for 15 minutes or so, and make the following adjustments:

1. With the throttle half-open, adjust the mixture screw for best rpm. The governor might "hunt," causing rpm to rise and fall.
2. Close the throttle to a fast (1600 rpm minimum) idle and rapidly flick it open. If the engine stumbles, back out the mixture screw an eighth of a turn and repeat the acceleration test.
3. The final adjustment can give an overly rich idle, but that is a design limitation not easily corrected.

Air filters

Improperly maintained air filters are engine-killers, as destructive as dirty crankcase oil. The assembly should make an airtight seal with the carburetor mouth and the filter element should be periodically cleaned. A dirty element creates a pressure drop, which encourages air leaks and particle infiltration.

Four types of filters are used. Probably the least effective are the gauze or fiber composition filters sometimes found on two-cycle engines. The fuel fog that hovers around the carburetor air horn keeps these filters wetted between service intervals.

Industrial engines are sometimes fitted with centrifugal oilbath filters, as shown in FIG. 3-22. While hardly state of the art, an oil bath filter can be effective if periodically cleaned in solvent and reoiled. Overfilling the reservoir will cause the engine to smoke.

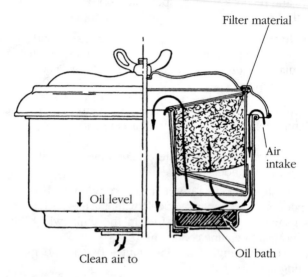

3-22 *Oil bath filter formerly used on Kohler engines. Heavier solids centrifuge out and fall into the oil reservoir; smaller particles must negotiate the mesh element, where most are trapped.*

Polyurethane, or sponge, filters are as efficient as any, but require frequent attention. Clean the element with solvent or hot water and detergent. Pour a few ccs of motor oil over the sponge, kneading it in gently. It is good—but messy—practice to reoil the filter each time the engine is started. Oil tends to migrate to the bottom of the unit.

Throwaway paper filters are the new four-cycle standard. No significant maintenance is possible, since exposure to oil or water swells the paper fibers to impermeability. Replacement elements are fairly expensive.

The best, most cost-effective filters, used on serious engines, combine a polyurethane outer element with a pleated-paper inner element (FIG. 3-23). The washable outer element

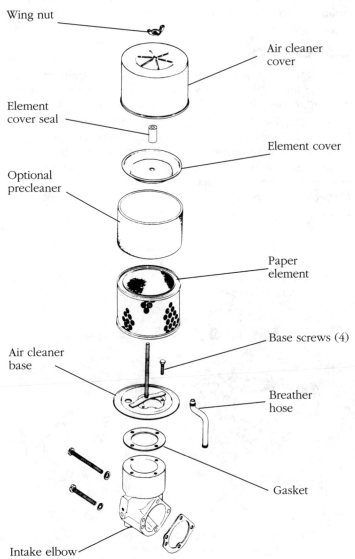

Wing nut

Air cleaner
cover

Element
cover seal

Element cover

Optional
precleaner

Paper
element

Base screws (4)

Air cleaner
base

Breather
hose

Gasket

Intake elbow

3-23 *Some engines are fitted with a two-stage air cleaner assembly, consisting of polyurethane precleaner and a paper inner element. When installing a new foam precleaner, turn it inside out to make a good seal with the inner element. Wisconsin application shown.*

should be serviced at least every 25 operating hours and more frequently in dusty environment.

Governors

Most small engines employ a governor to help maintain engine rpm under varying loads and to limit maximum speed. Air vane governors, such as the one shown in FIG. 3-24, sense engine speed as a function of cooling air pressure and velocity. The air vane is installed under the shroud, in the cooling airstream. It is spring-loaded and connects to the throttle, either directly as shown or by a linkage. As engine rpm increases, air pressure reacts on the vane to close the throttle; if speed drops below a certain value, the spring opens the throttle.

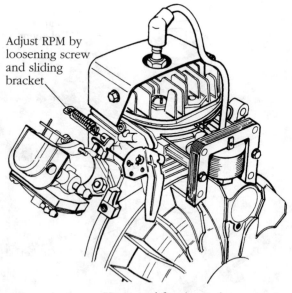

Adjust RPM by loosening screw and sliding bracket.

Horizontal fixed speed
(aluminum air vane governor)

3-24 *Air vane governor used on Tecumseh two-cycle engines. Engine runs at a preset speed, set by the spring bracket adjustment.*

Mechanical governors sense engine speed as flyweight movement (FIG. 3-25). Cam or crankshaft-driven flyweights respond to increasing engine speed by moving outward. This movement, acting through a spool and yoke, rotates the governor shaft and attached governor lever. The lever, working through a linkage, tends to close the throttle as rpm increases and to pull it open as rpm drops. Most mechanical governors allow the operator to vary engine speed between idle and allowable maximum.

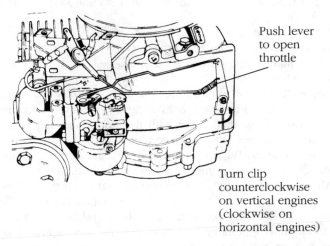

Push lever
to open
throttle

Turn clip
counterclockwise
on vertical engines
(clockwise on
horizontal engines)

3-25 *The relationship between the governor shaft and wide-open throttle is the crucial aspect of small engine service work. Unfortunately, procedures are not standardized (as this Tecumseh illustration shows).*

The more elaborate mechanisms incorporate a low-speed adjustment (distinct from the throttle stop screw) and can have provision to adjust governor sensitivity to speed variations. All governors allow the upper rpm limit to be varied by moving the governor arm relative to its shaft and the shaft relative to the yoke. No universal procedure applies; unless you understand the geometry of the linkage and of the flywheel mechanism (which is not visible), you can destroy the engine with this adjustment.

Warning: Do not attempt governor adjustments unless you have access to an accurate tachometer and factory literature for the engine in question.

Fuel pumps

If you are getting fuel to the pump, and no fuel out of it, the source of the difficulty is the pump diaphragm or check valves (which might be integral with the diaphragm). If you have no fuel to the pump, but fuel in the tank, the problem is a clogged tank screen or an air leak at a suction-side fuel line connection.

Mechanical pumps drive off the camshaft or crankshaft eccentric. Unlike automotive types, which they otherwise resemble, small engine pumps are rebuildable (FIG. 3-26). Procedures are fairly obvious, but a few points deserve mention:

Briggs & Stratton pumps are initially assembled with the pump lever spring installed, but disengaged from the actuating lever. Position the diaphragm with the slot in the diaphragm spindle 90 degrees from the lever. Rotate the diaphragm to capture the lever. With a screwdriver, pry the free end of the lever spring into engagement with the tang on the lever.

All mechanical pumps, of whatever manufacture, will be damaged if cover screws are tightened against a taut diaphragm. Stroke the lever to its travel limit and hold it in that position while tightening the screws. Torque screws in a criss-cross pattern in three increments.

Grease the end of the pump lever where it contacts the drive cam.

Use a new mounting gasket, coated with nonhardening Permatex or equivalent. Rotate the flywheel to relieve tension on the lever, and install the pump, pulling down the capscrews evenly.

Diaphragm pumps are generally simple one- or two-diaphragm affairs, as shown in FIG. 3-27. The more complex units, such as those used by Onan, employ mechanical check valves.

Scribe-mark the housings before disassembly.

Note the relationship between the diaphragm and gasket. Some gaskets mount on the outboard side of the diaphragm; some are under it. Clean metallic parts in solvent and replace "soft" parts.

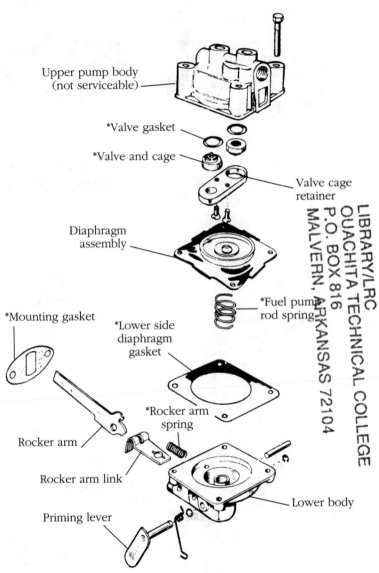

Upper pump body
(not serviceable)

*Valve gasket

*Valve and cage

Valve cage
retainer

Diaphragm
assembly

*Fuel pump
rod spring

*Mounting gasket

*Lower side
diaphragm
gasket

*Rocker arm
spring

Rocker arm

Rocker arm link

Lower body

Priming lever

LIBRARY/LRC
OUACHITA TECHNICAL COLLEGE
P.O. BOX 816
MALVERN, ARKANSAS 72104

*Parts included in repair kit.

3-26 *Onan mechanical fuel pump employs mechanical check valves that should pass air in one direction and block it in the other.*

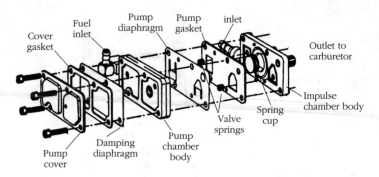

3-27 *Briggs & Stratton diaphragm-type fuel pump operates from crankcase pressure fluxuations.*

4

Rewind starters

Unlike other engine systems that operate continuously, manual and electric starters are designed for intermittent use, which is why rewind starters can get by with nylon bushings, and why motor pinions can cheerfully bang into engagement with the flywheel. The starter usually lasts about as long as the engine and the owner is satisfied.

But the balance between starter and engine life goes awry if the engine is allowed to remain chronically out of tune. Most starter failures are the result of overuse: The starter literally works itself to death cranking a bulky engine. Whenever you repair a starter, you must also—if the repair is to be permanent—correct whatever it is that makes the engine reluctant to start in the first place.

Side pull

The side-pull rewind (recoil, self-winding, or retractable) starter was introduced by Jacobsen in 1928 and has changed little in the interim. These basic components are always present:

- Pressed steel or aluminum housing, which contains the starter and positions it relative to the flywheel.
- Recoil spring, one end of which is anchored to the housing, the other to the sheave.

- Starter rope (nylon, although a few Fairbanks Morse wirelines are still encountered), which is anchored to and wound around the sheave.
- Sheave or pulley.
- Sheave bushing between sheave and housing or (on Briggs & Stratton) between sheave and crankshaft.
- Clutch assembly.

Troubleshooting

Most failures have painfully obvious causes, but it might be useful to have an idea of what you are getting into before the unit is disassembled.

Broken rope   The most common failure, often the result of putting excessive tension on the rope near the end of its stroke or by pulling the rope at an angle to the housing. The problem is exacerbated by a worn rope bushing (the guide tube, at the point where the rope exits the housing). In general, rope replacement means complete starter disassembly, although some designs allow replacement with the sheave still assembled to the housing.

Loss of spring tension   Usually the result of a broken spring, loss of tension can but also be caused by spring disengagement from the housing or sheave. The spring anchor slot on Briggs & Stratton housings can batter out and release the spring. Complete disassembly is required.

Rope fails to completely retract   On a unit in service, suspect loss of spring preload tension due to aging. The best recourse is to replace the spring, although preload tension can be increased by one sheave revolution. On a just-repaired unit, check the starter housing/flywheel alignment, spring preload tension, and replacement rope length and diameter.

Failure to engage flywheel   This failure is a clutch problem, caused by a worn or distorted brake spring (which can be a coil-type brake spring or a more vulnerable hair spring), retainer screw backout, or oil on clutch friction surfaces. While recoil spring and sheave bushing generally require some lubrication, starter clutch mechanisms must, as a rule, be assembled dry.

Excessive force required to pull rope   Check starter housing/flywheel alignment first. Then remove the starter, turn the engine over by hand to verify that it is free, and check starter action. The problem might involve a dry sheave bushing.

Noise from starter as engine runs   Check starter housing/flywheel alignment. On Briggs & Stratton designs, the problem is often caused by a dry sheave bushing (located between the starter clutch and crankshaft). Remove the blower housing and apply a few drops of oil to the crankshaft end.

Service procedures

Rewind starters are a special technology, and an overall view of the subject is helpful. The first order of business is to release spring preload tension, which can be done in two ways. Any rewind starter can be disarmed by removing the rope handle and allowing the sheave to unwind in a controlled fashion. Other starters have provision for tension release with the handle still attached to the rope. Briggs & Stratton provides clearance between sheave diameter and housing that allows several inches of rope to be fished out of the sheave groove. This increases the effective length of the rope, enabling sheave and attached spring to unwind. Many other designs incorporate a notch in the sheave for the same purpose (Fig. 4-1).

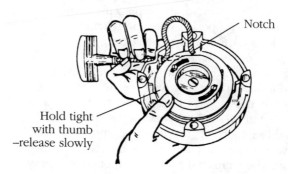

Notch

Hold tight
with thumb
–release slowly

4-1 *Common sense dictates that the starter should be disarmed before sheave is detached. Most have provision to unwind the rope a turn or so while others are disarmed by removing the rope handle and allowing the rope to fully retract.*

Brake the sheave with your thumbs as it unwinds. It is also good practice to number the sheave rotations from the point of full rope retraction so that the same preload can be applied on assembly.

The sheave is secured at its edges by crimped tabs and located by the crankshaft extension (Briggs & Stratton side pull), or else it rotates on a pin attached to the starter housing. A screw (Eaton) or retainer ring (Fairbanks-Morse, and several foreign makes) secures the sheave to the post.

The mainspring lives under the sheave, coiled between sheave and housing, with its inner, or movable, end secured to the sheave hub. The outer, or stationary, spring end anchors to the housing.

Warning: Even after preload tension is dissipated, rewind springs store energy that can erupt when the sheave is disengaged from the housing. Wear safety glasses.

The manner in which recoil springs secure to the housing varies among makes, and this affects service procedures. Some Eaton starters use an integral spring retainer that indexes to slots in the housing (FIG. 4-2). Spring and retainer are handled as a unit and should not be disassembled.

What might be called the "standard" attachment strategy is to secure the spring to a post pressed into the underside of the housing. The fixed end of the spring forms an eyelet or hook that slips over the anchor post. To simplify assembly, most manufacturers supply replacement springs coiled in a retainer clip. You position the spring and retainer in the housing cavity with the spring eyelet over the post and press the spring out of the retainer (which is then discarded). Sheave engagement usually takes care of itself. Exceptions are discussed in sections dealing with specific starters.

Some starters adapt to left- or right-hand rotation by reversing the spring (FIG. 4-3). Viewing the starter housing from the underside and using the movable spring end as reference, clockwise engine rotation demands counterclockwise spring windup. The wrap of the rope must, of course, provide appropriate sheave rotation.

The third type of spring anchor takes the form of a slot in the starter housing through which the spring passes. Figure 4-4 shows a Briggs & Stratton side-pull unit that is similar to several OMC Lawnboy types.

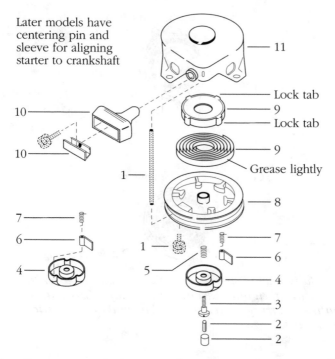

Later models have centering pin and sleeve for aligning starter to crankshaft

11

Lock tab
9
Lock tab

9
Grease lightly

8

4-2 *Eaton rewind starter with integral mainspring and housing that should not be dismantled in the field. These starters can be recognized by lock tabs on the spring housing outside diameter. This starter also uses a small coil spring—shown directly below the sheave—to generate friction on the clutch assembly.*

These starters are assembled by winding the spring home with the sheave. Thread the movable end of the spring through the housing slot, engage the movable end with the sheave, and rotate the sheave opposite engine rotation until the whole length of the spring snakes through the housing slot. The fixed end of the spring is notched or hooked for retention by the slot.

Rewind spring preload is necessary to maintain some rope tension when the rope is retracted. Too little preload and the rope handle droops; too much and the spring binds solid to pull out of its anchors.

Clockwise engine rotation

Cover

Rewind spring

Counterclockwise engine rotation

Cover

Rewind spring

4-3 *Many rewind springs and all ropes can be assembled for lefthand or righthand engine rotation. This feature is a manu-facturing convenience that makes life difficult for mechanics.*

Depending upon starter make and model, either of two approaches is used to establish preload. Most manufacturers suggest this general procedure:

1. Remove the rope handle if it is still attached.
2. Secure one end of the rope to the sheave anchor.
3. Wind the rope completely over the sheave, so that the sheave will rotate in the direction of engine rotation when the rope is pulled.
4. Wind the sheave against engine rotation a specified number of turns. If the specification is unknown, wind until the spring coil binds, then release the sheave for one or two revolutions.
5. Without allowing the sheave to unwind further, thread the rope through the guide tube (also called a ferrule, bushing, or eyelet) in the starter housing and attach the handle.
6. Gently pull the starter through to make certain that the rope extends to its full length before the onset of coil bind and that the rope retracts smartly.

Another technique can be used when the rope anchors to the inboard (engine) side of the sheave:

1. Assemble sheave and spring.
2. Rotate the sheave, winding the mainspring until coil bind occurs.
3. Release spring tension by one to no more than two sheave revolutions.
4. Block the sheave to hold spring tension. Some designs have provision for a nail that is inserted to lock the sheave to the housing; others can be snubbed with Vise-Grips or C-clamps.
5. With rope handle attached, thread rope through housing ferrule and anchor it to the sheave.
6. Release the sheave block and, using your thumbs for a brake, allow the sheave to rewind, pulling rope after it.
7. Test starter operation.

The starter rope should be the same weave, diameter, and length as the original. If required length is unknown, anchor the rope to the sheave, wind the sheave until coil bind—an operation that also winds the rope on sheave—then release the sheave for one or two turns, and cut the rope (leaving enough surplus for handle attachment).

Three types of clutch assemblies are encountered: Briggs & Stratton sprag, or rachet; Fairbanks-Morse friction-type; and the positive-engagement dog-type used by other manufacturers. In event of slippage, clean the Briggs clutch and replace the brake springs on the other types. Fairbanks-Morse clutch shoes can respond to sharpening.

One last general observation concerns starter positioning: Whenever a rewind starter has been removed from the engine or has vibrated loose, starter clutch/flywheel hub alignment must be reestablished. Follow this procedure:

1. Attach the starter or starter/blower housing assembly loosely to the engine.
2. Pull the starter handle out about 8 inches to engage the clutch.
3. Without releasing the handle, tighten the starter hold-down screws.
4. Cycle the starter a few times to check for possible clutch drag or rope bind. Reposition as necessary.

Briggs & Stratton

Briggs & Stratton side-pull starters are special in several respects (FIG. 4-4). In addition to its basic function of transmit-

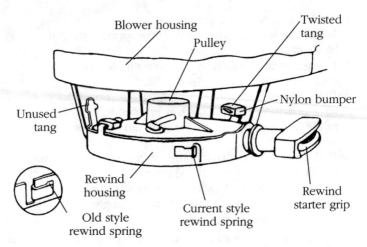

Blower housing
Pulley
Twisted tang
Nylon bumper
Unused tang
Rewind housing
Old style rewind spring
Current style rewind spring
Rewind starter grip

4-4 *Briggs & Stratton rewind starter used on 6 through 11 cubic inch engines. A later variant employs redesigned sheave and dispenses with nylon bumpers.*

ting torque from the starter sheave to the flywheel and disengaging when the engine catches, the starter clutch also serves as the flywheel nut and starter sheave shaft. Starter and blower housing assemblies are integral. It is possible, however, to drill out the spot welds and replace the starter assembly as a separate unit. Bend-over tabs locate the starter sheave in the starter housing.

Disassembly

Follow this procedure:

1. Remove blower housing and starter from engine.
2. Remove rope by cutting the knot at the starter sheave (visible from underside of blower housing).
3. Using pliers, grasp the protruding end of the mainspring and pull it out as far as possible (FIG. 4-5). Disengage the spring from the sheave by rotating the spring a quarter turn or by prying one of the tangs up and twisting the sheave.
4. Clean and inspect. Replace the rope if it is oil-soaked or frayed. Although it might appear possible to reform the end of a broken Briggs & Stratton mainspring, such efforts are in vain and the spring must be replaced for a

permanent repair. The same holds for the spring anchor slot in the housing. Once an anchor has swallowed a spring, the housing should be renewed.

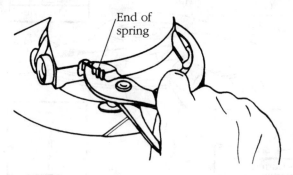

4-5 *Once rope is removed, pull the rewind spring out of the starter housing. The spring can be detached from sheave by twisting either sheave or spring a quarter turn.*

Assembly

1. Dab a spot of grease on the underside of the steel sheave. Note that a plastic version requires no lubrication. See FIG. 4-6.
2. Secure the blower housing engine side up to the workbench with nails or C-clamps.
3. Working from the outside of the blower housing, pass the inner end of the mainspring through the housing anchor slot. Engage the inner end with the sheave hub.
4. Some mechanics attach rope (less handle) to the sheave at this point. The rope end is cauterized in an open flame and is knotted.
5. Bend tabs to give the sheave $\frac{1}{16}$-inch endplay. Use nylon bushings on early models so equipped.
6. Using a $\frac{3}{4}$-inch wrench extension bar or a piece of one-by-one inserted into the sheave centerhole, wind the sheave 16 turns or so counterclockwise until the full length of the mainspring passes through the housing slot and the coil binds.
7. Release enough mainspring tension to align the rope anchor hole in the sheave with the housing eyelet.

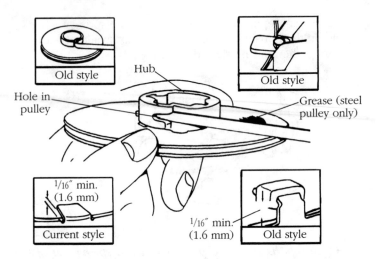

Old style

Hub

Hole in pulley

Grease (steel pulley only)

Old style

¹/₁₆″ min. (1.6 mm)

Current style

¹/₁₆″ min. (1.6 mm)

Old style

4-6 *Spring installation varies slightly with the date of manufacture. Steel sheaves require lubrication.*

8. Temporarily block the sheave to hold spring tension. You can block it with a crescent wrench snubbed between the winding tool and the blower housing. See A of FIG. 4-7.

9. If the rope has been installed, extract the end from between sheave flanges, thread through eyelet, cauterize, and attach handle. If the rope has not been installed, pass the cauterized end through the eyelet from outside the housing, between sheave flanges, and out through the sheave anchor hole (FIG. 4-7). Knot the end of the rope. Old-style sheaves incorporate a guide lug between flanges. The rope must pass between the lug and sheave hub. This operation is aided by a small screwdriver or a length of piano wire (A of FIG. 4-7).

The clutch is not normally opened unless it slips, a condition that can be caused by wear or the presence of lubricant. Old-style assemblies are secured with a wire retainer clip; newer models depend upon retainer-cover tension and can be pried apart with a small screwdriver (FIG. 4-8). Clean parts with a dry rag (avoiding use of solvent). The clutch housing can be removed from the crankshaft using a special factory wrench described in Chapter 3.

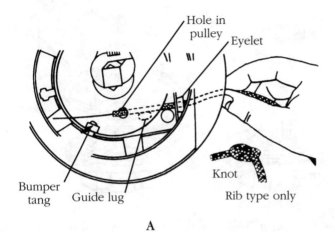

Hole in pulley

Eyelet

Bumper tang

Guide lug

Knot

Rib type only

A

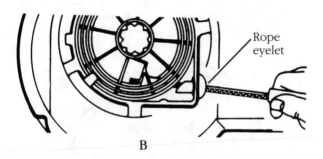

Rope eyelet

B

4-7 *Cast-iron block Briggs & Stratton engines never seem to wear out, and starters with internal rope guide lugs are still encountered. Use a length of piano wire to push the top past the inner side of the lug as shown (A). Newer designs omit the guide lug, making installation easier (B).*

Eaton

Recognizable by P-shaped engagement dogs, or pawls, Eaton starters are found on a wide variety of American engines. Light-duty models employ a single pawl; more substantial types use two, and high-torque models have three. Eaton pioneered the use of mainspring and spring retainer (a feature that makes life easier—and perhaps safer—for mechanics). Another Eaton feature is the centering pin, which it usually

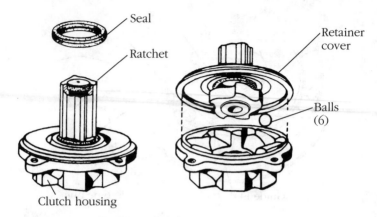

4-8 *Current production clutch cover is a snap fit to clutch housing. Older version employed a spring wire retainer. As a point of interest, older engines can be modified to accept new clutch assembly by trimming ⅜ inch from the crankshaft stub and ½ inch from sheave hub.*

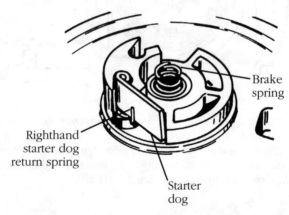

4-9 *Eaton rewind starter partially disassembled. Generous retainer screw torque compresses brake spring, generating friction against retainer that extends dog. Because the rope attaches to the inboard—and accessible—side of the sheave, the rope can be replaced by applying and holding mainspring pretension. Original rope is fished out, new rope is passed through the eyelet and sheave hole, knotted, and pretension is slowly released. Spring winds rope over sheave.*

rides on a nylon bushing. Figure 4-9 shows a single-pawl Eaton starter with a mainspring and centering pin.

The most common complaint is failure to engage the flywheel. This difficulty can be traced to the clutch brake, which generates friction that translates into pawl engagement, or to the pawls themselves. Two brake mechanisms are encountered. The latest arrangement, shown in FIG. 4-9, employs a small coil spring that reacts against the cup-like pawl retainer.

Another brake mechanism, used for many years and apparently still in production, interposes a star-shaped brake washer between the pawl retainer and brake spring. Figures 4-10 and 4-11 show this part. A shouldered retainer screw secures the assembly to the sheave and preloads the brake spring (FIG. 4-12).

Check the retainer screw, which should be just short of "hernia tight"; inspect friction parts, with special attention to the optional star brake; and check the pawl return spring (FIG. 4-9), which can be damaged by engine backfire. Clean parts, assemble without lubricant, and observe pawl response as the rope is pulled. If necessary, replace the star brake, retainer cup, and brake spring.

Figure 4-11 shows the top-of-the-line Eaton starter used on industrial engines. Service procedures are slightly more complex than for lighter-duty units because the sheave is split. This makes rope replacement more difficult, and the mainspring, which is not held captive in a retainer, can thrash about when the sheave is removed.

Disassembly

1. Remove the five screws securing the starter assembly to the blower housing.
2. Release spring preload. Most heavy-duty models employ a notched sheave that allows rope slack for disarming (see FIG. 4-2).
3. Remove the retainer screw and any washers that might be present.
4. Lift off the clutch assembly, together with the brake spring and the optional brake spring washer.
5. Carefully extract the sheave, keeping the mainspring confined within the starter housing.

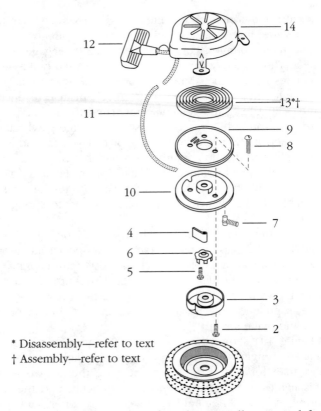

4-10 *Eaton light-duty pattern found on small two- and four-stroke engines. This starter is distinguished by its uncased mainspring (13) and single-dog clutch (dog shown at 4, clutch retainer at 3). In event of slippage during cranking, replace friction spring 5 and brake 6.*

Warning: Wear safety glasses during this and subsequent operations.

6. Remove the rope, which may be knotted on the inboard side of sheave or sandwiched between sheave halves as shown in FIG. 4-11. The screws that hold the sheave halves together can require a hammer impact tool to loosen.
7. Remove the spring if it is to be replaced. Springs without retainer are unwound one coil at a time from the center outward.
8. Clean and inspect with particular attention to the clutch

mechanism. Older light-duty and medium-duty models employed a shouldered clutch retainer screw with a 10-32 thread. This part can be updated to a 12-28 thread (Tecumseh PN. 590409A) by retapping the sheave pivot shaft.

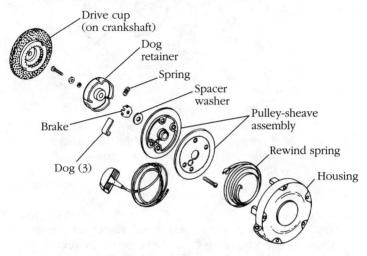

4-11 *Eaton heavy-duty starter of the type used on some Kohler engines. Note brake start washer, three-dog clutch, and split sheave.*

Assembly

1. Apply a light film of grease to the mainspring and sheave pivot shaft. Do not overlubricate, because the brake spring and clutch assembly must be dry to develop engagement friction. Snowproof clutches, recognizable by application and by half-moon pawl cam, might benefit from a few drops of oil on the pawl posts.
2. Install the rewind spring. Loose springs are supplied in a disposable retainer clip. Position the spring—observing correct engine rotation as shown in FIG. 4-3—over the housing anchor pin. Gently cut the tape holding the spring to the retainer, retrieving the tape in segments. Install spring and retainer sets by simply dropping them in place.
3. Install the rope, an operation that varies with sheave construction:

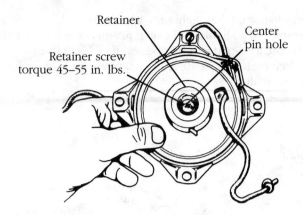

Retainer

Center
pin hole

Retainer screw
torque 45–55 in. lbs.

4-12 *View looking toward the inboard side of the sheave with dog assembled over return spring and brake spring in place. This unit is to be assembled dry; only snowproof models, distinguished by half-moon cam that engages the dogs, require oil on dog-mounting posts.*

Split sheave

A. Double-knot the rope, cauterize, and install between sheave halves, trapping the rope in the cavity provided.

B. Install the sheave on the sheave pivot shaft, engaging the inner end of the mainspring. A punch or piece of wire can be used to snag the spring end as shown in FIG. 4-13. Install the clutch assembly.

C. Wind the sheave until the mainspring coil binds (FIG. 4-14).

D. Carefully release spring tension two revolutions and align the rope end with the eyelet in the starter housing.

E. Using Vise-Grips, clamp the sheave to hold spring tension and guide the rope through the eyelet. Attach the handle.

F. Verify that sufficient pretension is present to retract rope.

One-piece sheave

A. Wind the sheave to coil bind and back off to align the rope hole on the inboard face of the sheave with the housing eyelet.

B. Clamp the sheave.

C. Cauterize the ends of the ropes and install the rope through the eyelet and sheave (FIG. 4-15).

D. Knot the rope under the sheave and install the handle.

E. Carefully release the sheave, allowing the rope to wind as the spring relaxes.

F. Test for proper pretension.

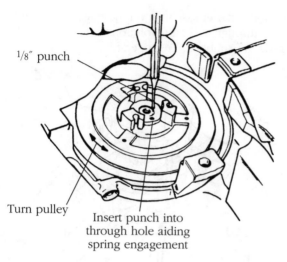

4-13 *A punch aids spring-to-sheave engagement on large Eaton starters.*

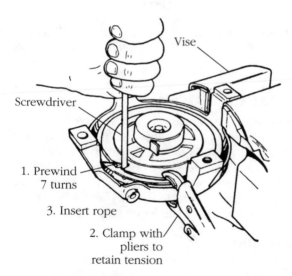

4-14 *Prewind specification varies with starter model and mainspring condition.*

4. Pull out the centering pin (where fitted) so that it protrudes about ⅛ inch past the end of the clutch retainer screw. Some models employ a centering pin bushing.

5. Install the starter assembly on the engine, pulling the starter through several revolutions before the hold-down screws are snubbed. Test operation.

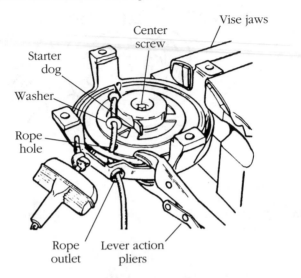

4-15 *Installing rope on one-piece sheave involves passing rope from outside starter housing, through eyelet, and into sheave connection point.*

Fairbanks-Morse

Fitted to several American engines, Fairbanks-Morse starters can be recognized by the absence of serrations on the flywheel cup. The cup is a soft aluminum casting, and friction shoes (clutch or brake shoes) are sharpened for purchase. Early models used a wireline in lieu of the rope. Figure 4-16 is a composite drawing of Models 425 and 475, which are intended for larger single-cylinder engines.

Disassembly

1. Remove the starter assembly from blower housing.
2. Turn the starter over on bench and, holding the large washer down with thumb pressure, remove the retainer ring that secures the sheave and clutch assembly (FIG. 4-17A).
3. Remove the washer, brake spring, and friction shoe assembly. Normally, the friction shoe assembly is not broken down further.

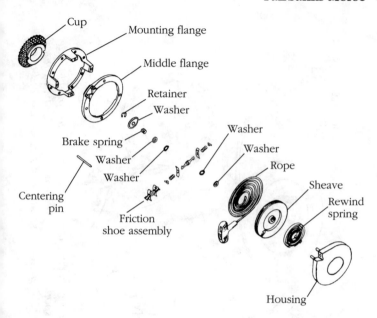

Cup
Mounting flange
Middle flange
Retainer
Washer
Washer
Washer
Brake spring
Washer
Rope
Washer
Sheave
Rewind
spring
Centering
pin
Friction
shoe assembly
Housing

4-16 *Fairbanks-Morse starter used on Kohler and other heavy-duty engines. Mounting and middle flanges are characteristic of F-M Model 475.*

4. Relieve mainspring preload by removing the rope handle and allowing the sheave to unwind in a controlled fashion. Tension on the Model 475 can be released by removing the screws holding the middle and mounting flanges together (B of FIG. 4-17).
5. *Cautiously* lift the sheave about ½ inch out of the housing and detach the inner spring end from the sheave hub.
6. Leave mainspring undisturbed (unless you are replacing it). To remove the spring, lift one coil at a time, working from the center outward. Wear eye protection.
7. Clean all parts in solvent and inspect.

Assembly

1. Install the spring, hooking the spring eyelet over the anchor pin on the cover. The spring lay shown in D of FIG. 4-17 is for conventional—clockwise when facing flywheel—engine rotation.
2. Rope installation and preload varies with starter model. In

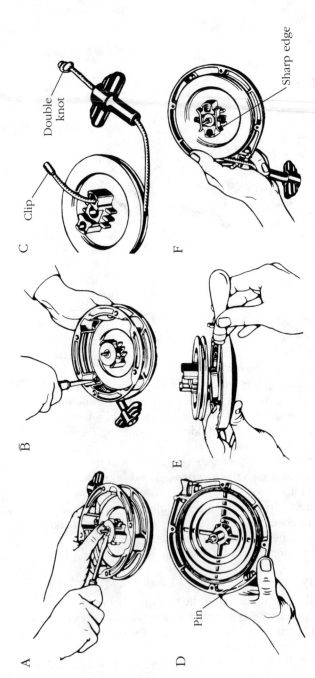

4-17 Crucial service operations and parts relationships include removing the retainer ring and spring-loaded washer (A), releasing residual spring tension, flanged-starter shown (B), rope anchors and rope lay, standard engine rotation (C), mainspring orientation, standard rotation (D), engaging spring and sheave (E), and correct brake shoe assembly (D).

all cases, the rope is attached to the sheave and wound on it before the sheave is fitted to the starter cover and mainspring. The Model 475 employs a split rope guide, or ferrule, consisting of a notch in the middle flange and in the starter housing. Consequently, the rope can be secured to and wound over the flange with the rope handle attached. Model 425 and most other Fairbanks-Morse starters use a one-piece ferrule and the rope must be installed without a handle. After the sheave is secured and the preload established, the rope is threaded through the ferrule for handle attachment.

3. Lubricate the sheave pivot shaft with light grease and apply a small quantity of motor oil to the mainspring. Avoid overlubrication.

4. Install the sheave over the sheave pivot shaft with the rope fully wound. With a screwdriver, hook the inner end of the spring into the sheave hub (E of FIG. 4-17).

5. Establish preload—four sheave revolutions against the direction of engine rotation for Model 425, five turns for Model 475, and variable for others.

6. Complete assembly, installing the sheave hold-down hardware and friction shoe assembly. When assembled correctly, the sharp edges of the friction shoes are poised for leading contact with the flywheel hub inside diameter (F of FIG. 4-17).

7. Pull the centering pin out about ⅛ inch for positive engagement with the crankshaft centerhole.

8. Install the assembled starter on the blower housing, rotating the flywheel with the starter rope as the hold-down screws are torqued. This procedure helps to center the clutch in the flywheel hub.

9. Start the engine to verify starter operation.

The Fairbanks-Morse utility starter is a smaller and simpler version of the heavy-duty models just discussed. The starter housing mounts directly to the engine cooling shroud (eliminating flanges). A one-piece sheave is used with the rope, anchored by a knot rather than a compression fitting. The utility starter uses the same clutch components as its larger counterparts and, like them, can be assembled for right-hand or left-hand engine rotation (See FIG. 4-18).

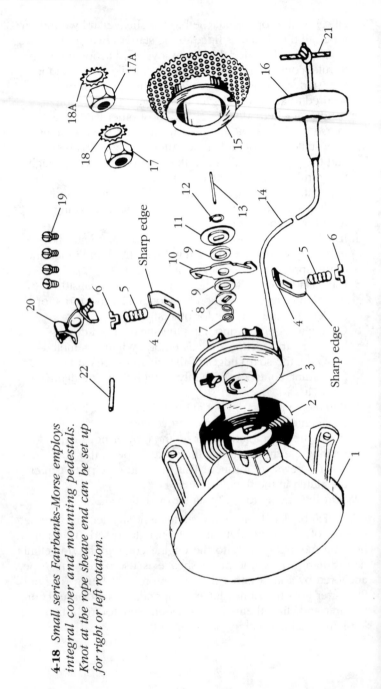

4-18 Small series Fairbanks-Morse employs integral cover and mounting pedestals. Knot at the rope sheave end can be set up for right or left rotation.

ILLUS. NO.	QTY.	DESCRIPTION
1	1	Cover
2	1	Rewind spring
3	1	Rotor
4	2	Friction shoe plate
5	2	Friction shoe spring
6	2	Spring retainer plate
7	1	Brake spring
8	1	Brake washer
9	2	Fiber washer
10	1	Brake lever
11	1	Brake retainer washer
12	1	Retainer ring
13	1	Centering pin
14	1	Cord
15	1	Cup and screen
16	1	T-handle
17	1	L.H. thick hex nut
17A	1	R.H. thick hex nut
18	1	Ext. tooth lockwasher (left hand)
18A	1	Ext. tooth lockwasher (right hand)
19	4	Pan hd. screw w/int.-ext. tooth lockwasher
20	1	Friction shoe assembly, includes: Items 4, 5, 6 and 10
21	1	Spiral pin
22	1	Roll pin

Vertical pull

Like other spring-powered devices, vertical-pull starters must be disarmed before disassembly. Otherwise, the starter will disarm itself with unpredictable results. Disarming involves three distinct steps: releasing mainspring pretension (usually by slipping a foot or so of rope out of the sheave flange and allowing the sheave to unwind), disengaging the mainspring anchor (usually held by a threaded fastener) and, when the spring is to be replaced, uncoiling the spring from its housing.

Warning: Safety glasses are mandatory for disassembly.

Vertical-pull starters tend to be mechanically complex and—because of a heavy reliance upon plastic, light-gauge steel, and spring wire—are unforgiving. Parts easily bend or break. Lay components out on the bench in proper orientation and in sequence of disassembly. If there is any likelihood of confusion, make sketches to guide assembly. Also note that the step-by-step instructions in this book must aim at thoroughness and describe all operations, but it will rarely be necessary to follow every step and completely dismantle a starter.

Briggs & Stratton

Briggs & Stratton uses one vertical-pull starter with minor variations in the link and sheave mechanisms. It is probably the most reliable of these starters, and the easiest to repair.

Disassembly

1. Remove starter assembly from engine.
2. Release mainspring pretension by lifting the rope out of the sheave flange and, using the rope for purchase, winding the sheave counterclockwise two or three revolutions (FIG. 4-19).
3. Carefully pry the plastic cover off with a screwdriver. Do not pull on the rope with the cover off and spring anchor attached; under these conditions it is possible for the outer end of the spring to slip out of the housing.
4. Remove the spring anchor bolt and spring anchor (FIG. 4-20). If the mainspring is to be replaced, carefully extract it from the housing, working from the center coil outward. Note the spring lay for future reference.

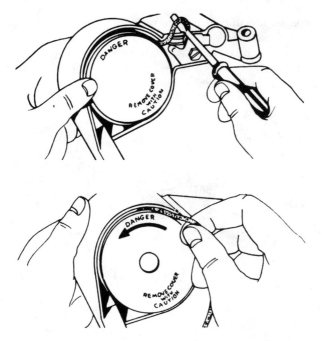

4-19 *Briggs & Stratton vertical-pull starters are disarmed by slipping rope out of sheave groove and using the rope to turn the sheave two or three revolutions counterclockwise until the mainspring relaxes.*

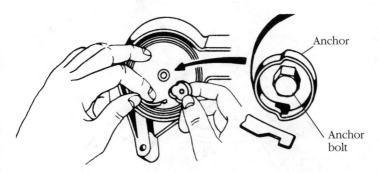

4-20 *Mainspring anchor bolt must be torqued 75–90 pounds-inch and can be further secured with thread adhesive.*

5. Separate the sheave and the pin (FIG. 4-21). Observe the link orientation.
6. The rope can be detached from the sheave with the aid of long-nosed pliers. Figure 4-22 shows this operation and link retainer variations.

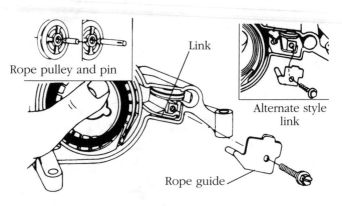

Rope pulley and pin

Link

Alternate style link

Rope guide

4-21 *Observe friction link orientation for assembly.*

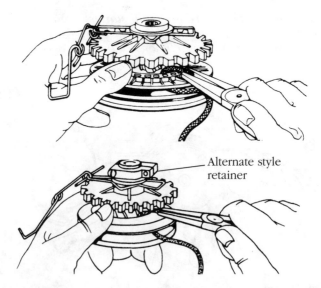

Alternate style retainer

4-22 *Rope can be disengaged from sheave with long-nosed pliers.*

7. The rope can be disengaged from the handle by prying the handle center section free and cutting the knot (FIG. 4-23).
8. Clean all parts (except rope) in petroleum-based solvent to remove all traces of lubricant.
9. Verify the gear response to link movement as shown in FIG. 4-24. The gear should move easily between its travel limits. Replace the link as necessary.

Insert

Grip

4-23 *Briggs & Stratton handle insert must be pried out of grip for rope installation.*

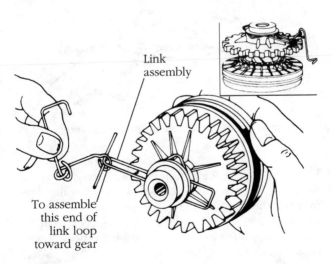

Link assembly

To assemble this end of link loop toward gear

4-24 *Pinion gear should move through its full range of travel in response to link movement. Note orientation of link for assembly (inset).*

Assembly

1. Install the outer end of mainspring in the housing retainer slot and wind counterclockwise (FIG. 4-25).
2. Mount the sheave, sheave pin, and link assembly in the housing. Index the end of the link in the groove or hole provided (FIGS. 4-26, 4-27).

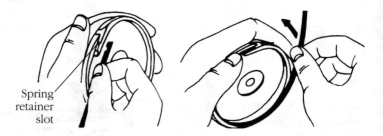

4-25 *Mainspring winds counterclockwise from outer coil.*

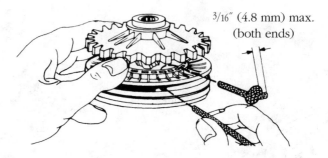

4-26 *A short length of piano wire aids rope insertion into sheave.*

3. Install the rope guide and hold-down screw.
4. Rotate the sheave counterclockwise, winding the rope over the sheave (FIG. 4-28).
5. Engage the inner end of the mainspring on the spring anchor. Mount the anchor and torque the hold-down capscrew to 75–90 pounds-inch.
6. Snap the plastic cover into place over the spring cavity.
7. Disengage 12 inches or so of rope from the sheave and, using rope for purchase, turn the sheave two or three revolutions clockwise to generate pretension. See FIG. 4-29.
8. Mount the starter on the engine and test.

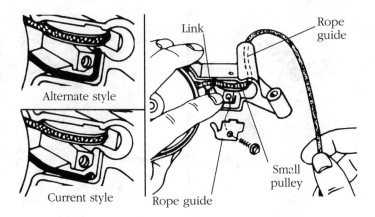

4-27 *Friction link hold-down detail.*

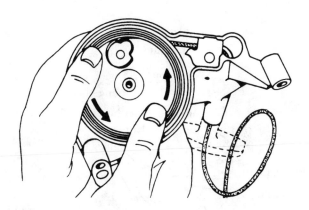

4-28 *The rope winds counterclockwise on the sheave, then the spring anchor and anchor bolt are installed.*

Tecumseh

Tecumseh has used several vertical-pull starters, ranging from quickie adaptations of side-pull designs in the 1960s to the current vertical-engagement type, which stands as a kind of textbook example of modern engineering and manufacturing techniques.

The *gear-driven starter* shown in FIG. 4-30 is an interesting transition from side to vertical-pull. No special service instructions seem appropriate, except to provide plenty of grease in

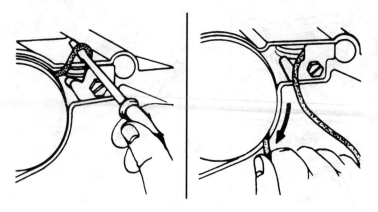

4-29 *Pretension requires two or three sheave revolutions using the rope for leverage.*

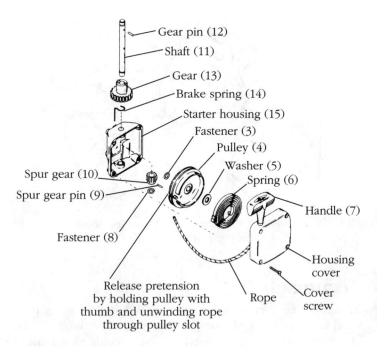

Gear pin (12)
Shaft (11)
Gear (13)
Brake spring (14)
Starter housing (15)
Fastener (3)
Pulley (4)
Washer (5)
Spring (6)
Spur gear (10)
Spur gear pin (9)
Handle (7)
Fastener (8)
Housing cover
Release pretension by holding pulley with thumb and unwinding rope through pulley slot
Rope
Cover screw

4-30 *Early Tecumseh vertical-pull starter, driving through a gear train. While heavy (and, no doubt, expensive to manufacture), this starter was quite reliable.*

the gear housing and some light lubrication on the mainspring. Assemble the brake spring without lubricant.

The current *horizontal-engagement starter* (FIG. 4-31) is reminiscent of the Briggs & Stratton design, with rope clip, cup-type spring anchor ("hub" in the drawing), and threaded sheave extension upon which the pinion rides.

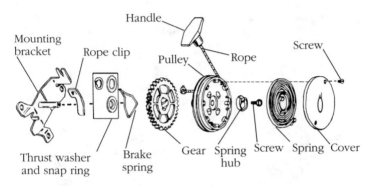

4-31 *Tecumseh's most widely used vertical-pull starter employs a spiral gear to translate the pinion horizontally into contact with the flywheel.*

Disassembly

1. Remove the unit from the engine.
2. Detach the handle and allow the rope to retract past the rope clip. This operation relieves mainspring preload tension.
3. Remove the two screws, while holding the cover in place, and carefully pry the cover free.
4. Remove the hold-down screw and hub.
5. Extract the mainspring from the housing, working a coil at a time from the center out. If it will be reused, the spring can be left undisturbed.
6. Lift off the gear and pulley assembly. Disengage the gear and, if necessary, remove the rope from the pulley.
7. Clean all parts except rope.
8. Inspect the friction spring (the Achilles' heel of vertical-pull starters). The spring must be in solid contact with the groove in the gear.

Assembly

1. Secure the rope to the handle, using No. 4½ or 5 nylon
   rope, 61 inches long for standard starter configurations.
   Sear the rope ends and form by wiping with a cloth while
   the rope is still hot.
2. Assemble the gear on the pulley, using no lubricant.
3. Lightly grease the center shaft and install the gear and
   pulley. The brake spring loop is secured by the bracket
   tab. The rope clip indexes with the hole in the bracket
   (FIG. 4-32).

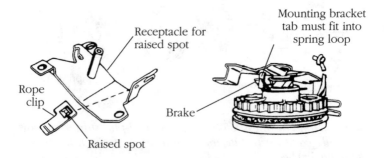

Receptacle for
raised spot

Mounting bracket
tab must fit into
spring loop

Rope
clip

Brake

Raised spot

4-32 *Generous gear lash, minimum ¹⁄₁₆ inch, is required to
assure pinion disengagement when engine starts.*

4. Install the hub and torque the center screw to 44–55
   pounds-inch.
5. Install the spring. New springs are packed in a retainer
   clip to make installation easier.
6. Install the cover and cover screws.
7. Wind the rope on the pulley by slipping it past the rope
   clip. When fully wound, turn the pulley two additional
   revolutions for preload.
8. Mount the starter on the engine, adjusting the bracket for
   minimum ¹⁄₁₆-inch tooth clearance (FIG. 4-33). Less
   clearance could prevent disengagement, destroying
   the starter.

Vertical pull, vertical engagement

The *vertical-pull, vertical-engagement* starter is a serious piece
of work that demands special service procedures. It is relative-

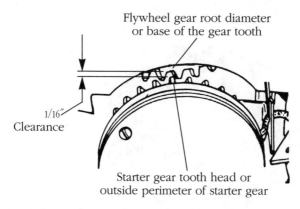

Flywheel gear root diameter
or base of the gear tooth

1/16″
Clearance

Starter gear tooth head or
outside perimeter of starter gear

4-33 *Rope clip and spring loop index to bracket.*

ly easy to disassemble while still armed. The results of this error can be painful. Another point to note is that rope-to-sheave assembly as done in the field varies from the original factory assembly.

Figure 4-34 is a composite drawing of several vertical-pull starters. Many do not contain the asterisked parts, and early models do not have the V-shaped groove on the upper edge of the bracket which simplifies rope replacement.

When this groove is present, the rope (No. 4½, 65-inch standard length, longer with remote rope handle) can be renewed by turning the sheave until the staple, which holds the rope to the sheave, is visible at the groove (FIG. 4-35). Pry out the staple and wind the sheave tight. Release the sheave just some 180 degrees to index the hole in the sheave with the V-groove. Insert one end of the replacement rope through the hole, out through the bracket. Cauterize and knot the short end, and pull the rope through, burying the knot in the sheave cavity. Install the rope handle, replacing the original staple with a knot, and release the sheave. The rope should wind itself into place.

Disassembly

1. Remove the starter from the engine.
2. Pull out the rope far enough to secure it in the V-wedge on the bracket end. This part, distinguished from the V-groove mentioned above, is called out in FIG. 4-34.

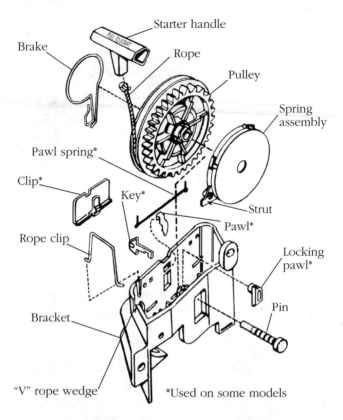

Starter handle

Brake

Rope

Pulley

Spring assembly

Pawl spring*

Clip*

Key*

Strut

Pawl*

Rope clip

Locking pawl*

Bracket

Pin

"V" rope wedge

*Used on some models

4-34 *Tecumseh's vertical-pull, vertical-engagement starter is the most sophisticated unit used on small engines. Spring and cover are integral and are not separated for service.*

3. The rope handle can be removed by prying out the staple with a small screwdriver.
4. Press out the head pin that supports the sheave and spring the capsule in the bracket. This removal can be done in a vise with a large deepwell wrench socket as backup.
5. Turn the spring capsule to align with the brake spring legs. Insert a nail or short (¾-inch maximum) pin through the hole in the strut and into the gear teeth (FIG. 4-36).
6. Lift the sheave assembly and spring the capsule out of the bracket.

Warning: Do not separate the sheave assembly and spring capsule until the mainspring is completely disarmed.

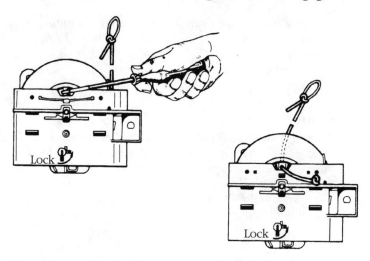

4-35 *V-groove in bracket gives access to rope anchor on some models.*

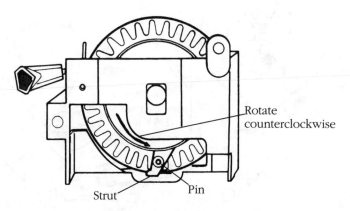

4-36 *A pin locks the spring capsule and gear to prevent sudden release of mainspring tension.*

7. Hold the spring capsule firmly against the outer edge of the sheave with thumb pressure and extract the locking pin inserted in Step 5.

8. Relax pressure on the spring capsule, allowing the capsule to rotate, dissipating residual mainspring tension.

9. Separate the capsule from the sheave and, if rope replacement is in order, then remove the hold-down staple from the sheave.

10. Clean and inspect all parts.

Note: No lubricant is used on any part of this starter.

Assembly

1. Cauterize and form ends of the replacement rope (see specs above) by wiping down with a rag while still hot.

2. Insert one end of rope into sheave, 180 degrees away from the original (staple) mount (A of FIG. 4-37).

3. Tie a knot and pull the rope into the knot cavity.

4. Install the handle (B of FIG. 4-37).

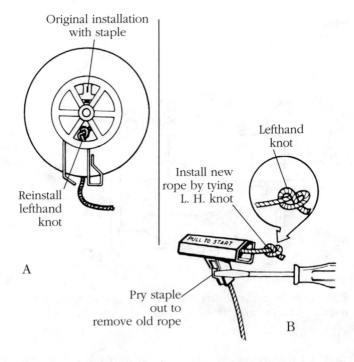

Original installation with staple

Reinstall lefthand knot

A

Lefthand knot

Install new rope by tying L. H. knot

Pry staple out to remove old rope

PULL TO START

B

4-37 *Replacement rope anchors with knot, rather than staple, and mounts 180 degrees from original position on sheave.*

5. Wind the rope clockwise (as viewed from the gear) on the sheave.
6. Install the brake spring, spreading the spring ends no more than necessary.
7. Position the spring capsule on the sheave, making certain the mainspring end engages the gear hub (A of FIG. 4-38).

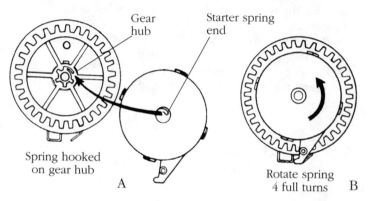

4-38 *Spring capsule engages gear hub (A), is rotated four revolutions and pinned (B).*

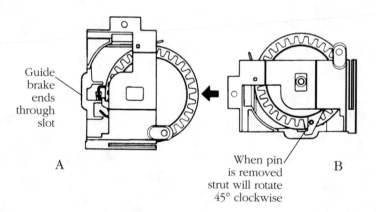

4-39 *Sheave and spring capsule assembly installs in bracket with brake spring ends in slots (A). Releasing pin arms starter (B), which can now be mounted on the engine.*

8. Wind four revolutions, align the brake spring ends with the strut (B of FIG. 4-38), and lock with the pin used during disassembly.
9. Install pawls, springs, and other hardware that might be present.
10. Insert the sheave and spring assembly into the bracket, with the brake spring legs in the bracket slots (FIG. 4-39).
11. Feed the rope under the guide and snub it in the V-notch.
12. Remove the locking pin, allowing the strut to rotate clockwise until retained by bracket.
13. Press or drive the center pin home.
14. Mount the starter on the engine and test.

5

Electrical system

At its most developed, the electrical system consists of a charging circuit, a storage battery, and a starting circuit. A flywheel alternator provides electrical energy that is collected in the battery for eventual consumption by the starter motor.

Not all small engine electrical systems include both circuits. Some dispense with the starting circuit and others employ a starting circuit without provision for onboard power generation.

Starting circuits

Starting circuits fall into two major groups: dc (direct current) systems that receive power from a 6 or 12V battery and ac (alternating current) systems that feed from an external 120Vac line. I will not discuss ac systems because the hazards implicit in line-current devices cannot be adequately addressed in a book of this type. My discussion is limited to dc systems that employ conventional (lead-acid) or nicad batteries.

Lead-acid

As shown in FIGS. 5-1 and 5-2, a conventional starting circuit includes four major components—battery, ignition switch,

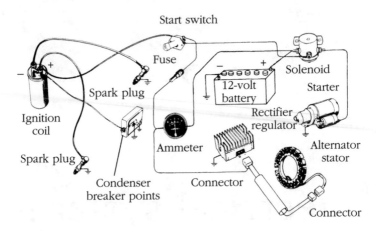

5-1 *Electrical system supplied with Onan engines and tied into battery-and-coil ignition. Note the heavy-duty stator and combined rectifier/regulator.*

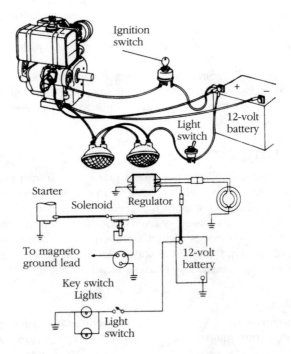

5-2 *Briggs & Stratton 10A system shows tie-in to charging system on positive battery post. Ignition switch grounds magneto and does not interchange with automotive-type switches.*

solenoid, and motor—wired into two circuit loops:

- Control loop—14 gauge primary wire from the positive battery terminal, through the ignition switch, to the solenoid field windings.
- Power loop—cable from the positive battery terminal, through the solenoid and to the starter motor.

The solenoid—more properly called a relay—is a normally open (NO) electromagnetic switch. When energized by the starter switch, the solenoid closes with an audible click to complete the power circuit. Most solenoids are internally grounded as shown, which means that mounting faces must be clean and hold-down bolts secure. Interlocks are sometimes included in the control circuit to prevent starting under unsafe conditions (e.g., if the machine is in gear). As shown in FIG. 5-2, the charging system connects to the positive post of the battery.

Figure 5-3 outlines diagnostic procedures, which use the solenoid as the point of entry. Shunting the solenoid with jumper cables cuts the solenoid and the control loop out of the circuit and gives an immediate indication of starter motor function.

Most starting circuit problems are the fault of the battery.

1. Remove the cable connections at the battery and scrape the battery terminals to bright metal.
   Repeat the process at the solenoid and starter connections.
2. Verify that the electrolyte covers the battery plates. Add distilled water as necessary, but do not overfill.
3. Clean the battery top with a mixture of baking soda, detergent, and water. Rinse with fresh water and wipe dry.
4. With the battery cables still disconnected, charge the battery.

Warning: Lead-acid batteries give off hydrogen gas during charging. To minimize sparking and possible explosion, connect the charger cables (red to positive, black to negative) before switching ON the charger; switch OFF the charger before disconnecting.

5. Test each cell with a hydrometer. Replace the battery if the charger cannot raise average cell readings to at least 1.260 or if individual cell readings vary more than 0.050.
6. Connect a voltmeter across the battery terminals. With ignition output grounded, crank the engine for a few seconds. Cranking voltage should remain above 9.5V (12V systems) or 4.5V (6V). Lower readings mean a defective battery or starter circuit.

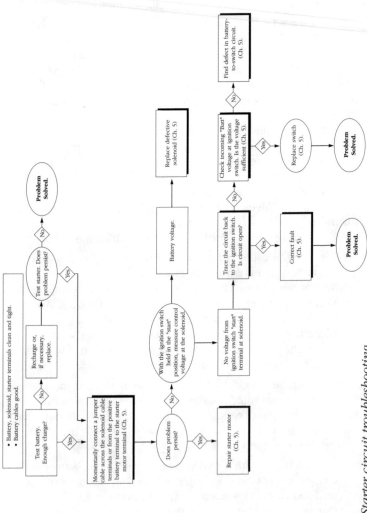

5-3 *Starter circuit troubleshooting.*

Caution: Small-engine starters have limited duty cycles. Allow several minutes for the starter to cool between 10-second cranking periods.

Nicad

As far as I am aware, only two manufacturers—Briggs & Stratton and Tecumseh Products—supply nicad-powered systems. As shown in FIG. 5-4, the circuit includes a nickel-cadmium battery pack, switch, and 12Vdc starter motor, all specific to these systems and not interchangeable with any other. A 110Vac charger replenishes the battery pack before use.

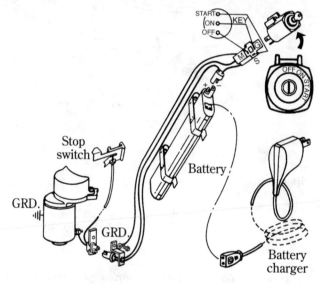

5-4 *Briggs & Stratton nicad system integrates engine controls on wiring harness. Tecumseh is electrically similar.*

Troubleshooting Diagnostic procedures are straightforward:

1. Verify the ability of the battery pack to hold a charge. If necessary, test the 110Vac charger.
2. If a known-good battery pack does not function on the engine, check the control switch with an ohmmeter.
3. Test the starter motor.

Battery pack Refusal to hold a charge is the most common fault. After 16 hours on the charger, battery potential should range between 15.5V and 18V. Assuming that voltages fall within these limits, the next step is to test capacity through a controlled discharge. If a carbon-pile tester is not available, connect two No. 4001 headlamp bulbs in parallel, as shown in FIG. 5-5. A freshly charged battery pack should illuminate the lamps brightly for 5 minutes (Briggs & Stratton) or 6 minutes (Tecumseh), figures that represent enough energy to start the average engine about 30 times.

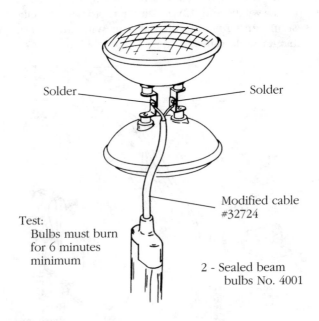

Solder

Solder

Modified cable
#32724

Test:
  Bulbs must burn
  for 6 minutes
  minimum

2 - Sealed beam
    bulbs No. 4001

5-5 *Nicad battery pack test load is fabricated from two sealed-beam headlamps and a battery-to-starter cable.*

Warning: Dispose of nicad battery packs in a manner approved by local authorities. Cadmium, visible as a white powder on leaking cells, is a persistent poison. Do not incinerate, and avoid welding on or around the battery pack.

Battery life will be extended if charging is limited to 12 or 16 hours immediately before use, and to once every two months in dead storage.

Nicad charger Output varies with battery condition, but after two or three hours, it should be about 80 mA. Tecumseh lists a test meter (PN 670235) for the PN 32659 charger; Briggs suggests that the technician construct a tester, as described in FIG. 5-6.

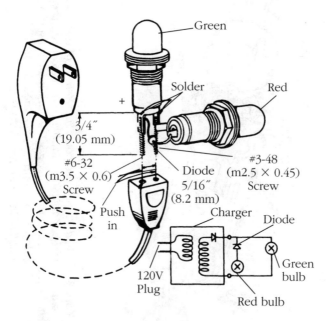

5-6 *A functional charger will light the green lamp only. A charger with an open diode will light the red bulb; one with a shorted diode will light both bulbs. Parts required: one 1N4005 diode, two Dialco lamp sockets (PN 0931-102 red, PN 0932-102 green), two No. 53 bulbs, one 6-32 ¾-inch screw and one 3-48 ¾-inch screw.*

Starter motors

Industrial motors, such as the Prestolite unit shown in FIG. 5-7, are rebuildable and can be serviced by most automotive electrical shops. The Briggs & Stratton motor, shown in FIG. 5-8, is also rebuildable (thanks to adequate parts support) and can, in the context of small engines, be considered a heavy-duty motor. American Bosch, used by Kohler and shown in FIG. 5-9, is, from a repairability point of view, on a par with the Briggs. European Bosch, Bendix, Nippon Denso, and Mitsubishi starter motors are of similar quality. Light-duty units, such as the nicad starter shown in FIG. 5-10, do not justify serious repair efforts.

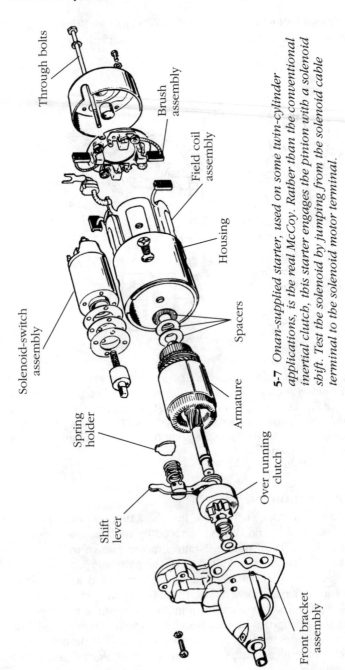

Through bolts

Brush assembly

Field coil assembly

Housing

Solenoid-switch assembly

Spacers

Armature

Spring holder

Over running clutch

Shift lever

Front bracket assembly

5-7 Onan-supplied starter, used on some twin-cylinder applications, is the real McCoy. Rather than the conventional inertial clutch, this starter engages the pinion with a solenoid shift. Test the solenoid by jumping from the solenoid cable terminal to the solenoid motor terminal.

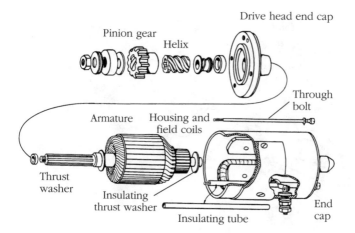

5-8 *Briggs & Stratton 12Vdc starter motor employs electro-magnetic (EM) fields, thrust washer on drive end and insulating thrust washer at commutator.*

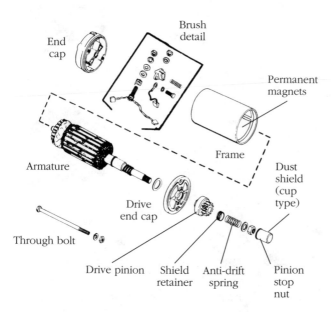

5-9 *American Bosch 12Vdc starter features permanent magnet (PM) fields and radial commutator with brushes parallel to motor shaft. Found on Kohler and other serious engines.*

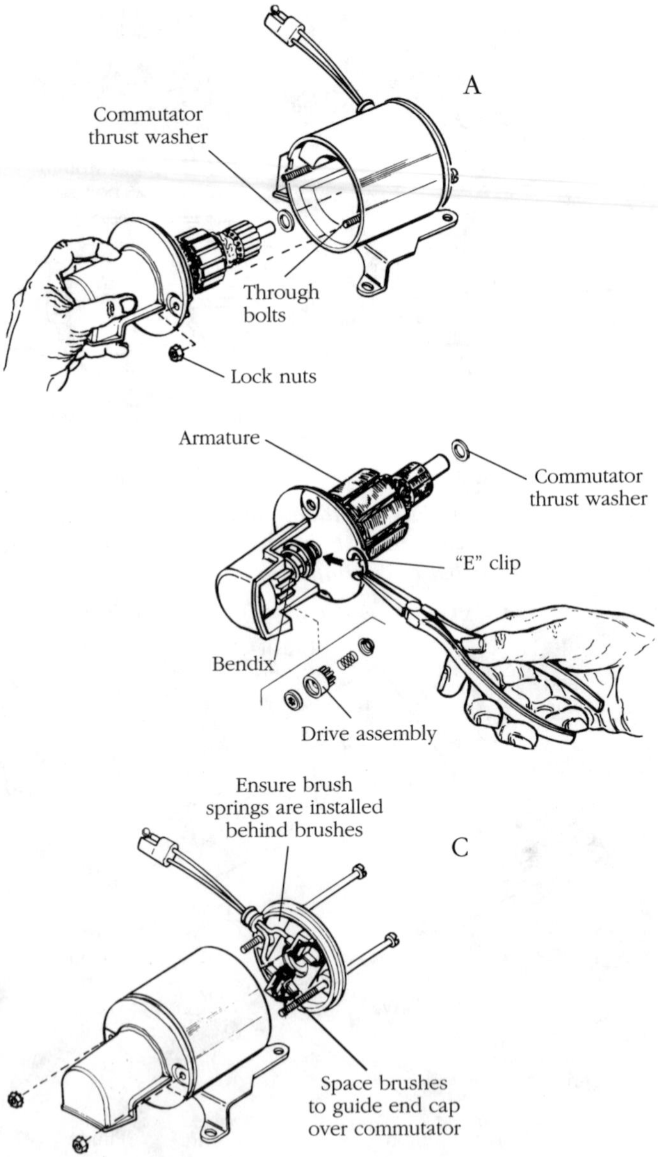

A

Commutator thrust washer

Through bolts

Lock nuts

B

Armature

Commutator thrust washer

"E" clip

Bendix

Drive assembly

C

Ensure brush springs are installed behind brushes

Space brushes to guide end cap over commutator

5-10 *Tecumseh nicad end caps assemble with through bolts (A). Bendix inertial clutch secures with an E-clip (B). Note the thrust washer. (C) Brushes, replaceable only as part of the cap assembly, must be shoehorned over the commutator. This particular starter should draw 20A while turning the engine 415 rpm (lube oil at 70°F).*

Troubleshooting   Figure 5-11 describes motor failure modes and likely causes. References to field failures apply only to those motors that use wound field coils; PM (permanent magnet) fields are, of course, immune to electrical malfunction.

It is assumed that full battery voltage reaches the starter and that the flywheel offers only normal resistance to turning.

Starter does not function:

- Brushes stuck in holders
- Dirty, oily brushes/commutator
- Open, internally shorted, or grounded field coil
- Open, internally shorted, or grounded armature

Starter cranks slowly
(minimal acceptable cranking speed = 350 rpm):

- Worn brushes or weak brush springs
- Dirty, oily, or worn commutator
- Worn shaft bushings
- Defective armature

Starter stalls under compression:

- Overly advanced ignition timing
- Defective armature
- Defective field coil

Starter works intermittently:

- Sticking brushes
- Loose connections in external circuit
- Dirty, oily commutator

Starter spins freely without turning flywheel:

- Pinion gear sticking on shaft
- Broken pinion and/or flywheel teeth

5-11 *Starter motor faults and probable causes.*

Repairs   With the exception of replacing the inertial clutch, repair procedures discussed here apply to heavy-duty motors.

Upon disassembly, clean the interior of the starter with an aerosol product intended for this purpose. Do not use a petroleum-based solvent. Note the placement of thrust and insulating washers.

- Inertial clutch—Shown clearly in FIG. 5-10 and tangentially in other drawings, the Bendix is serviced as a complete assembly. It secures to the motor shaft with a nut, spring clip, or roll pin. Support the free end of the motor shaft when driving the pin in or out. The helix and gear install dry, without lubrication; the pinion ratchet can be lightly oiled.
- End cap—Scribe the end cap and motor frame as an assembly aid. Installation can be tricky when the brush assembly is part of the cap. In some cases, you can retract the brushes with a small screwdriver. Radially deployed brushes can be retained with a fabricated bracket (FIG. 5-12).

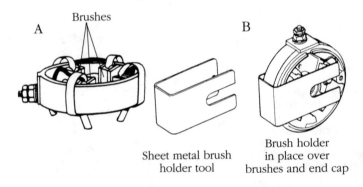

5-12 *Fabricated brush holders for four-pole (brush) assemblies (A) and two-pole (B).*

- Bushings—Do not disturb the bushings unless replacements are at hand. Drive out the pinion-end bushing with a punch sized to the bushing outside diameter. Commutator bushings must be lifted out of their blind end-cap bosses, which can be done by filling the cavity with grease and using a punch, sized to match shaft outside diameter as a piston. Hammer the punch into the grease.

    Sintered bronze bushings—Recognizable by their dull, sponge-like appearance, these bushings should be submerged in motor oil for a few minutes before installation. Brass bushings require a light, temperature-resistant grease, such as Lubriplate.
- Brushes—Most starter problems originate with brushes that wear short and bind against the sides of their holders. As a

rule of thumb, replace the brushes when worn to half their original length. Older starters used screw-type brush terminals; newer starters employ silver solder or integrate brushes and brush holders with the cap. Note the lay of the brushes: Rubbing surfaces must conform to the convexity of the commutator.

• Commutator—Heavy-duty starters can be "skimmed" on a lathe to restore commutator concentricity and surface finish. A light-duty commutator might benefit from light polishing with 000-grade sandpaper. Do not use emery cloth.

• Armature—Check continuity with an ohmmeter or 12V trouble light. Two conditions must be met:

~ paired commutator bars, under adjacent brush holders, have continuity.

~ no pair of bars has continuity with other pairs or with the motor shaft (FIG. 5-13).

A growler will detect internal shorts.

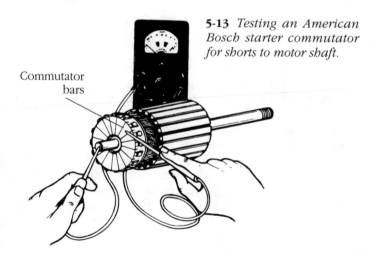

Commutator bars

5-13 *Testing an American Bosch starter commutator for shorts to motor shaft.*

• Fields—Inspect PM fields for mechanical damage (from contact with the armature) and for failure of the adhesive backing. Electromagnetic fields should be tested by a specialist, although at this point you have reached the end of practical repairability for even the best starter motor.

Charging circuits

In its most vestigial form, a charging circuit consists of a coil, flywheel magnet, and a load, such as a headlamp. Coil output alternates, or changes direction, each time a flywheel magnet excites it (the discussion of magneto theory in chapter 2 explains why). Voltage is speed-sensitive: At idle the lamp barely glows; at wide-open throttle the filament verges on self-destruct.

Adding a battery means that stator output must be rectified, or converted from ac to dc. This is almost always done by means of one or two silicon diodes, which act as check valves to pass current flowing in one direction and block it in the other. Single-diode rectifiers pass that half of stator output that flows in the favored direction (FIG. 5-14A). Full-wave rectifiers use two diodes, wired in a bridge circuit, to impose unidirectionality upon all of the output, so that none of it goes to waste (FIG. 5-14B).

The battery receives a charge so long as its terminal voltage is lower than rectifier output voltage. The battery also acts as a ballast resistor, limiting output voltage and current. Even so, these values remain closely tied to engine speed and not to electrical loads.

More sophisticated circuits use a solid-state regulator to synchronize charging current and voltage with battery requirements. The regulator caps voltage output at about 14.7V and responds to low battery terminal voltage with more current.

The usual practice is to encapsulate the regulator with the rectifier. Look for a potted "black box" or a finned aluminum can, mounted under or on the engine shroud. All of these units share engine ground with the stator and battery. Hold-down bolts must be secure and mating surfaces clean.

Most regulator/rectifiers have three wires going to them, as shown in FIGS. 5-1, 5-2, and 5-15. Two of these wires carry ac from the stator and one conveys B+ voltage to the battery. Wires that supply B+ voltage are often, but not always, color-coded red. When in doubt, check wire colors at the stator connections.

Twenty and 30A systems can include a bucking coil (a kind of electrical brake) to limit output. The presence of such a coil is signaled by one (or sometimes two) additional wires from the stator to regulator-rectifier.

Use a high-impedance meter, preferably digital, for voltage checks. Identify circuits before testing, with particular atten-

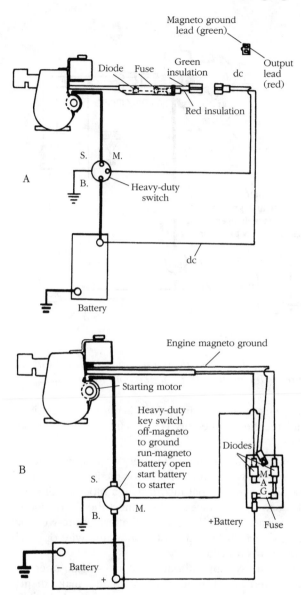

5-14 *Tecumseh 3A systems illustrate two approaches to rectification. Single-diode, half-wave rectifier, located in wiring harness, passes half of stator output to battery (A). Two-diode, full-wave rectifier utilizes all of stator output, doubling the charge rate (B). In event of overcharging, one diode can be removed.*

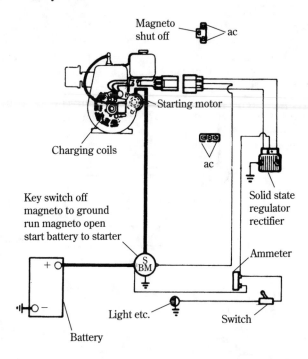

Magneto shut off · ac

Starting motor

Charging coils

ac

Key switch off
magneto to ground
run magneto open
start battery to starter

Solid state
regulator
rectifier

Ammeter

+

S
BM

Light etc.

Switch

–

Battery

5-15 *Most regulator/rectifiers have four connections: ac, ac, B+, and engine ground through the hold-down bolts.*

tion to the magneto primary, which is often integrated into the regulator-rectifier connector. That circuit carries some 300Vac.
    Do not:

• Reverse polarity —reversed battery or jumper-cable connections will ruin the regulator/rectifier on all but the handful of systems that incorporate a blocking diode.
• Introduce stray voltages —disconnect the B+ rectifier-to-battery lead before charging the battery or arc-welding.
• Create direct shorts —do not ground any wire or touch ac output leads together.
• Operate the system without a battery —when open-circuit, unregulated ac output tests are permitted, make them quickly at the lowest possible engine rpm/voltage needed to prove the stator.
• Run the engine without the shroud in place —if necessary, route test leads outside of the shroud.

Figure 5-16 presents a standard troubleshooting format used by many small engine mechanics. It applies to all unregulated systems and to more than 90 percent of systems with a regulator or regulator-rectifier. There are exceptions: Certain regulators and regulator/rectifiers do not tolerate hot (engine-running) disconnects. These components will be damaged by attempts to measure open-circuit ac voltage.

An extensive inquiry has uncovered two of these maverick systems; there are almost certainly others among the thousands of models and types of small engines sold in the U.S. (Kawasaki, a relatively minor player in this market, lists more than 300 distinct models).

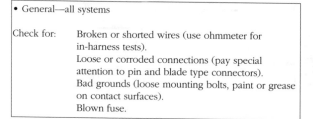

ac generating circuits-w/o battery or regulator, 6 or 12 volt

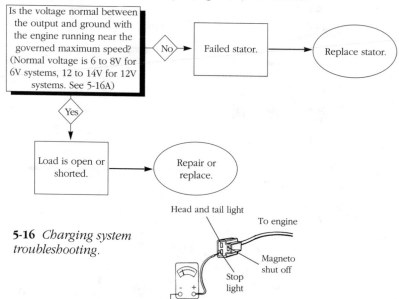

5-16 *Charging system troubleshooting.*

5-16 *Continued.*

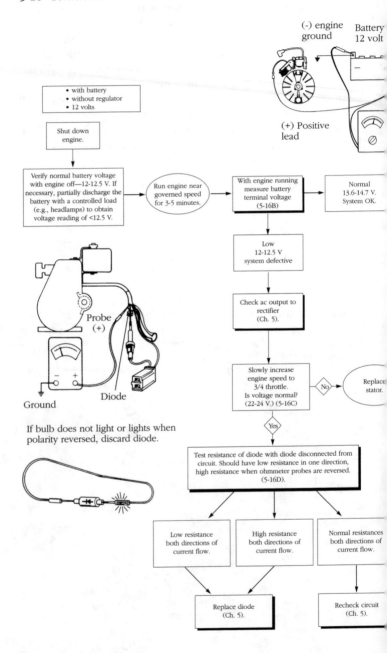

(-) engine ground　Battery 12 volt

• with battery
• without regulator
• 12 volts

Shut down engine.

(+) Positive lead

Verify normal battery voltage with engine off—12-12.5 V. If necessary, partially discharge the battery with a controlled load (e.g., headlamps) to obtain voltage reading of <12.5 V.

Run engine near governed speed for 3-5 minutes.

With engine running measure battery terminal voltage (5-16B)

Normal 13.6-14.7 V. System OK.

Low 12-12.5 V system defective

Check ac output to rectifier (Ch. 5).

Probe (+)

Ground　Diode

If bulb does not light or lights when polarity reversed, discard diode.

Slowly increase engine speed to 3/4 throttle. Is voltage normal? (22-24 V.) (5-16C)

No　Replace stator.

Yes

Test resistance of diode with diode disconnected from circuit. Should have low resistance in one direction, high resistance when ohmmeter probes are reversed. (5-16D).

Low resistance both directions of current flow.

High resistance both directions of current flow.

Normal resistances both directions of current flow.

Replace diode (Ch. 5).

Recheck circuit (Ch. 5).

Charging circuits with regulator/rectifier

EXCEPT: • Tecumseh 7A
 • Syncro 20A
 • Others that do not tolerate battery or ac
   output–lead disconnects. Refer to text (Ch. 5).
PRECONDITIONS: Good, well-charged system battery.

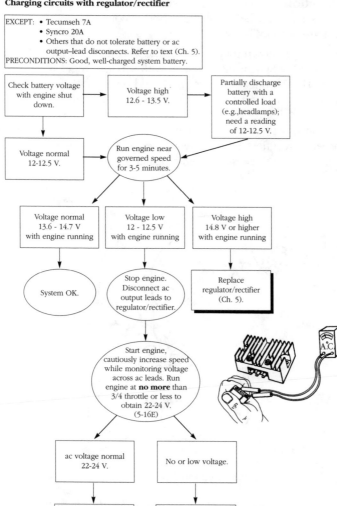

Caution: Contact a factory-trained mechanic or a manufacturer's tech rep before making hot disconnects on any system you are not familiar with.

The two known mavericks are the Tecumseh-supplied 7A system and the Synchro 20A system with separate regulator and rectifier. The Tecumseh 7A system, found on some 3- to 10-hp side valve engines and the overhead valve OMV 120, uses any of three under-shroud regulator-rectifiers shown in FIG. 5-17; the caption describes the ac voltage test procedure. Synchro regulators and rectifiers are clearly labeled with their manufacturer's name. Return these systems to a dealer for service.

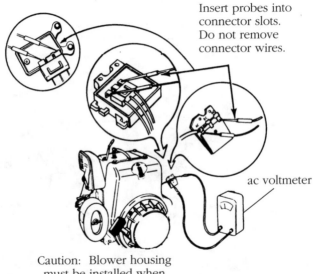

Insert probes into connector slots. Do not remove connector wires.

ac voltmeter

Caution: Blower housing must be installed when running engine.

5-17 *Tecumseh 7A system cannot tolerate open-circuit ac voltage measurements. Test ac output as shown, with regulator/rectifier electrically connected and the engine cooling shroud in place. (Unlike other rectifier/regulators, these units do not require an engine ground.) Minimum acceptable stator performance is: 16.0Vdc at 2500 rpm; 19.0Vac at 3000 rpm; 21.0VAC at 3300 rpm.*

6

Engine mechanical

Numerical data about the reliability of their products is not something that engine makers like to talk about, but it is known that Kohler's top-line overhead valve engines carry a D-10 rating of 1500. That is, 90 percent of these engines continue to run for 1500 hours—or about 70 days—on a dynamometer programmed to replicate real-world loads and speeds. To judge from Kohler's almost legendary reputation, these numbers probably represent the upper limit for air-cooled industrial engines.

The 10 percent of engines that stop running during the test do so because of parts failure. The D-10 rating measures consistency of performance, or what engineers call reliability. But some engines seem immune to parts breakage and, like very old men, simply wear out. Measurement of resistance to wear, or durability, would require something on the order of a D-90 test.

Reliability depends upon the design of the parts, the materials used, and the quality control exercised by the manufacturer. Durability is a function of the above, plus tolerance stackup. Parts are manufactured in a range of tolerances. For example, Tecumseh HH120 cylinder bores range from 3.500 to 3.501 inch; pistons for the same engine have diameters of between 3.495 and 3.497 inch. Piston clearance thus ranges from 0.006 to 0.003 inch, depending upon the luck of assembly.

Preventive maintenance (PM) will minimize, if not entirely eliminate, reliability problems and make some contribution to increased durability. Effective PM has three elements: careful record-keeping, scheduled service, and a willingness to intervene at the first sign of less-than-perfect performance. Records contain the history of the engine and of the maintenance effort, keyed to date and, if possible, operating hours.

A notebook is adequate for one or two small engines; keeping abreast of a fleet of engines requires a computer and the appropriate software. Programs are available that can automatically schedule maintenance tasks, keep track of replacement parts in inventory and on back order, and rate these parts in terms of cost effectiveness.

Routine tasks—changing oil and filters, decarbonizing combustion chambers, retorquing head bolts, cleaning cylinder fins, replacing point and condenser sets—follow a schedule, usually keyed to engine operating hours. (But an engine that remains idle for long periods also needs routine attention.) Although it is difficult to break out these matters precisely, it appears that the routine performance of such tasks has the most effect on durability.

The third element of the program addresses reliability. Mechanics must be taught to be proactive—that is, to intervene early, before catastrophic failures develop. Obviously some common sense must be exercised, but the controlling idea is for the mechanic to seek out work rather than wait for something to break.

This approach to maintenance has proved itself many times. EMD diesel engines were giving 20,000 hours between overhauls for a Houston-based drilling contractor, which was consistent with the manufacturer's expectations. Detailed record-keeping, rigid scheduling, and a proactive style of maintenance increased engine life to 30,000 hours with no increase in unscheduled breakdowns.

Troubleshooting

Chapter 1 describes the quick checks (compression, bearing tightness, and crankshaft straightness) that should be made as a kind of reconnaissance before the engine is disassembled. TABLE 6-1 lists the more common engine malfunctions and their probable causes. Figure 6-1 details why four-cycle engines develop a thirst for oil; FIG. 6-2 explains where the power goes.

Table 6-1. Engine related malfunctions (assuming ignition, fuel, and starting systems properly adjusted and working).

Symptom	Probable causes
Crankshaft locked	Jammed starter drive
	Hydraulic lock—oil or raw fuel in chamber
	Rust-bound rings (cast-iron bores only)
	Bent crankshaft
	Parted connecting rod
	Broken camshaft
Crankshaft drags when turned by hand	Bent crankshaft
	Lubrication failure, associated with cylinder bore and/or connecting rod
Crankshaft alternately binds and releases during cranking (rewind or electric starter)	Bent crankshaft
	Incorrect valve timing
	Loose blade/blade adapter (rotary lawnmower)
	Loose, misaligned flywheel
No or weak cylinder compression	Blown head gasket
	Leaking valves
	Worn cylinder bore/piston/piston rings
	Broken rings
	Holed piston
	Parted connecting rod
	Incorrect valve timing
No or imperceptible crankcase compression (two-cycle)	Leaking crankcase signals
	Leaking crankcase gaskets
	Failed reed valve (engines so equipped)
Rough, erratic idle	Stuck breather valve
	Leaking valves

Table 6-1. Continued

Symptom	Probable causes
Misfire, stumble under load	Improper valve clearance
	Weak valve springs
	Leaking carburetor flange gasket
	Leaking crankcase seals (two-cycle)
Loss of power	Loss of compression
	Leaking valves
	Incorrect valve timing
	Restricted exhaust ports/muffler (two-cycle)
	Leaking crankcase seals (two-cycle)
Excessive oil consumption (four-cycle)	Faulty breather
	Worn valve guides
	Worn or glazed cylinder bore
	Worn piston rings/ring grooves
	Worn piston/cylinder bore
	Clogged oil-drain holes in piston
	Leaking oil seals
Engine knocks	Carbon buildup in combustion chamber
	Loose or worn connecting rod
	Loose flywheel
	Worn cylinder bore/piston
	Worn main bearings
	Worn piston pin
	Excessive crankshaft endplay
	Excessive camshaft endplay
	Piston reversed (engines with offset piston pins)
	Loose PTO adapter
Excessive vibration	Loose or broken engine mounts
	Bent crankshaft

Oil leaks at crank-shaft seals	Hardened or worn seal
	Scored crankshaft
	Bent crankshaft
	Worn main bearings
	Scored oil seal bore allowing oil to leak around seal outside diameter
	Seal tilted in bore
	Seal seated too deeply in bore blocking oil return hole
	Breather valve stuck closed
Crankcase breather passes oil (four-cycle)	Leaking gasket
	Dirty or failed breather
	Clogged drain hole in breather box
	Piston ring gaps aligned
	Leaking crankshaft oil seals
	Valve cover gasket leaking (overhead valve engines)
	Worn rings/cylinder/piston

Loss of crankcase compression in two-cycle engines can be difficult to diagnose. The piston acts as pump, pressurizing the crankcase to force the air-fuel mixture through the transfer port and into the chamber. Pressures are quite low (5 or 6 psi during cranking) and difficult to detect as resistance to fly-wheel movement.

Even so, experienced mechanics can sense the absence of compression as the engine is slowly pulled through. Failure of the crankcase to hold pressure is almost always the fault of one or both crankshaft seals.

Partial seal failure is almost impossible to diagnose from the flywheel. Meaningful symptoms are loss of power from a lean mixture (seals almost always leak in both directions, diluting the fuel charge with air) and refusal of the engine to run unless the choke is engaged.

You can test crankcase integrity, although the effort is hardly worthwhile. You must fabricate rubber-gasketed blocking plates for the carburetor intake and exhaust. One plate includes a Schraeder fitting, connected to a squeeze bulb and a pressure gauge. To test, immerse the engine in a tub of solvent and

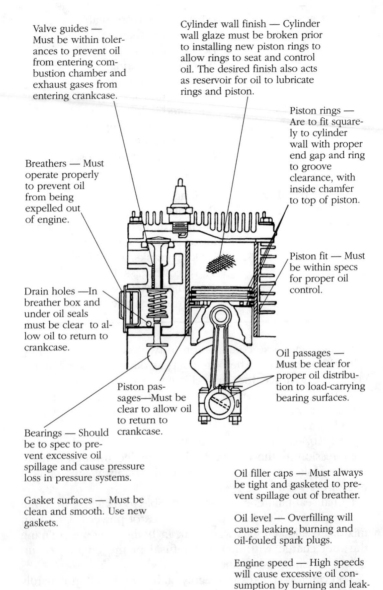

Valve guides — Must be within tolerances to prevent oil from entering combustion chamber and exhaust gases from entering crankcase.

Cylinder wall finish — Cylinder wall glaze must be broken prior to installing new piston rings to allow rings to seat and control oil. The desired finish also acts as reservoir for oil to lubricate rings and piston.

Piston rings — Are to fit squarely to cylinder wall with proper end gap and ring to groove clearance, with inside chamfer to top of piston.

Breathers — Must operate properly to prevent oil from being expelled out of engine.

Piston fit — Must be within specs for proper oil control.

Drain holes —In breather box and under oil seals must be clear to allow oil to return to crankcase.

Oil passages — Must be clear for proper oil distribution to load-carrying bearing surfaces.

Piston passages—Must be clear to allow oil to return to crankcase.

Bearings — Should be to spec to prevent excessive oil spillage and cause pressure loss in pressure systems.

Gasket surfaces — Must be clean and smooth. Use new gaskets.

Oil filler caps — Must always be tight and gasketed to prevent spillage out of breather.

Oil level — Overfilling will cause leaking, burning and oil-fouled spark plugs.

Engine speed — High speeds will cause excessive oil consumption by burning and leaking.

6-1 *Factors that affect oil consumption (four-cycle engines).*

Ignition — Must be properly timed so that spark plug fires at precise moment for full power.

Cylinder head — Should not be warped. Gasket surface must be true.

Valves — Check for seating, warping, sticking. Grind and lap to proper angle.

Cylinder head bolts — Tighten to proper torque.

Valve seats — Must be of specified angle and width.

Spark plug gap — Adjust to proper setting, use round feeler guage.

Cylinder head gasket — Must form perfect seal between cylinder and head.

Valve guide — Examine for wear, varnish which may prevent proper valve action.

Fins — Keep clean to prevent power loss because of overheating.

Valve springs — Check free length, must have proper tension to close valve and hold on seat.

Piston rings — Piston rings must be fitted properly with recommended end gap to ensure sufficient pressure on cylinder wall to transfer heat and seal high pressure.

Valve gaps — Must be adjusted properly.

Cam lobes — Check for wear, must be proper size to open valve fully to allow complete discharge of exhaust and intake of fuel.

Piston pin — Must allow friction free movement of connecting rod and piston.

Piston fit — Must be fitted to cylinder with recommended clearance.

Oil passages — All oil holes and passages must be clear to allow full lubrication for friction free operation.

Connecting rod — Match marks must be matched and connecting rod nuts tightened to proper torque.

Air filter — Should be clean to allow engine to breath.

Carburetor — Must be set properly to assure proper and sufficient air and fuel.

6-2 *Factors that affect power output (four-cycle engines).*

pressurize the crankcase to about 5 psi. Seal leaks show as bubbles around the crankshaft.

General procedure

Make haste slowly when working on an unfamiliar engine. Carefully note (and write down if necessary) the orientation of the parts, the lengths and washer configuration of fasteners, and the alignment of timing marks. Wear-matched parts should be assembled with their mates. Plastic zipper bags help keep things in order. Clean parts twice—once upon removal as a prelude to inspection, and again just prior to oiling and assembly. Use a torque wrench on critical fasteners, such as connecting-rod and cylinder-head bolts.

Cylinder head

Most utility and industrial engines employ demountable cylinder heads, sealed with throwaway composition gaskets and secured to the block by capscrews.

Warning: Composition gaskets employ asbestos as a filler. Dispose of the gasket in a safe manner. Carefully scrape gasket remains from the block and head without breathing or ingesting *any* dust that is generated.

With the engine at room temperature, remove the capscrews. Note variations in length, such as those found on aluminum-block Briggs & Stratton engines. Some European makes employ studs that pass through or around the cylinder barrel and anchor in the crankcase. These engines can be equipped with reusable copper head gaskets. Anneal the gasket by heating it with a propane torch and quenching it in oil or water.

Remove carbon deposits from the cylinder head, piston top, and block. An end-cutting wire brush is the preferred tool, although a dull knife can also be used on stubborn deposits. Be careful not to gouge the aluminum or damage the gasket surfaces.

Inspect the spark plug boss for stripped or pulled threads. Repairs can be made with a 14mm Heli-Coil kit. Check head distortion with the aid of a surface plate or piece of plate (not window) glass. The head is considered acceptable if a 0.003-inch feeler gauge will not pass between bolt holes (FIG. 6-3).

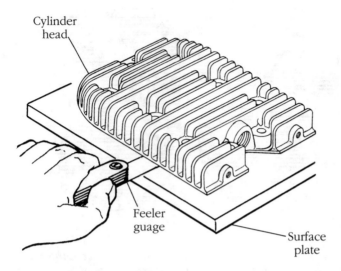

Cylinder head

Feeler guage

Surface plate

6-3 *Cylinder head flatness should be checked to assure gasket integrity. Commercial plate glass can be substituted for the surface plate shown.* Kohler

Ideally, a warped head should be replaced, along with the head bolts. If the engine is fitted with a single head casting, however, the head can be reground without serious side effects.

Separate head castings, such as found on horizontally opposed twin-cylinder engines, can also be ground, but great care must be exercised to take off equal amounts of metal on both. Tape a piece of medium-grit wet-or-dry emery paper to the plate glass and, applying pressure at a point near the center of the head, grind the gasket surface. Oil speeds the process. When the gasket surface is uniformly bright, the head is flat.

Install a new gasket and torque in three equal increments to specification. Normally, cylinder head bolts are lubricated with motor oil. The torque sequence for four-bolt heads is a simple X pattern. Others are torqued from the center bolts outward so that the ends of the casting go down last. However, the factory might make exception to this general rule (as shown in FIG. 6-4). Consult your factory manual for the engine in question.

Valves

Side valves—i.e., those located in the block—are removed and installed with either of the compressor tools shown in FIG. 6-5.

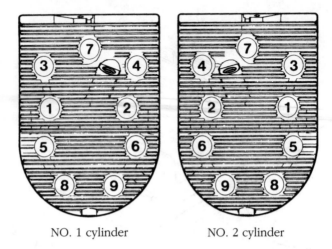

NO. 1 cylinder   NO. 2 cylinder

6-4 *Cylinder head torque sequence varies with engine make and model. Onan B43E, 43G, and 48G patterns differ between cylinders.*

Rotate the flywheel to seat the valve, insert the compressor under the valve collar, compress the valve spring, and remove the valve locks. It is good practice to temporarily plug the oil drain hole in the valve chamber floor to prevent a valve lock from falling in the crankcase.

Split locks are almost universal; Tecumseh and, most notably, Briggs & Stratton use cross pins. Some Briggs models employ one-piece retainers. (FIG. 6-6). Installation is the reverse of disassembly. Split locks can be positioned with the aid of grease and a screwdriver as (shown in B of FIG. 6-5). Professional mechanics generally prefer to use a magnetic insertion tool such as Snap-On's CF 771.

When properly secured, split locks are swallowed by the collar and are no longer visible. Briggs' one-piece retainer is centered under the collar and the cross pin is tucked out of sight under the collar.

A side-valve engine can be serviced without special tools (although the procedure costs something in frustration). Lift the collar with two flat screwdrivers. The trick is to keep the collar level so the valve remains seated as the spring compresses. Split locks should fall or can be knocked free; other types are removed with long-nosed pliers. Installation is a bit more difficult, particularly if you are working alone. Compress

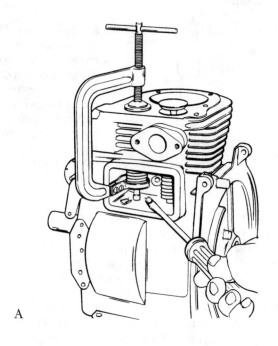

A

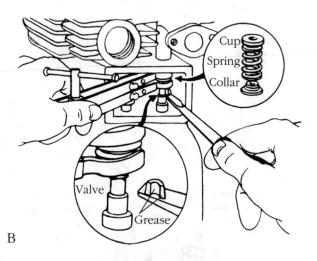

B

6-5 *Use a clamp (A) or bridge-type (B) spring compressor to remove and install block-mounted valves. The former tool is available from Kohler, the latter from Briggs & Stratton. Note how split valve locks are spooned into place with the aid of a grease-coated screwdriver.*

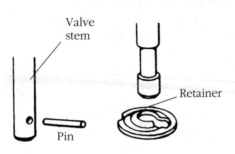

Valve
stem

Retainer

Pin

6-6 *In addition to split keepers (the current norm, shown in Figs. 6-5 and 6-7, you might encounter pin locks and slotted retainers. Both of these mechanisms im-pose high unit loads and wear fairly rapidly.*

the spring as before, and make certain that the valve is seated. Hold pressure on the screwdrivers with one hand and insert the valve locks. Split keepers might slip partly out of the valve stem groove, but can be tapped into place against spring tension.

Overhead valve (OHV) mechanisms are more accessible and consequently easier to service than side-valve units. Detach the cylinder head and support the head and valves on a wood block sized to fit the combustion cavity. The purpose of the block is to prevent valve movement as springs are compressed. Collapse the springs with a crow's-foot tool (FIG. 6-7).

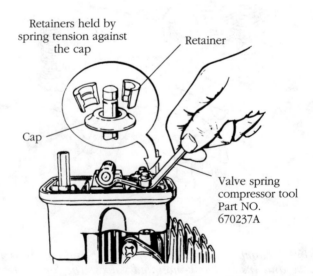

Retainers held by
spring tension against
the cap

Retainer

Cap

Valve spring
compressor tool
Part NO.
670237A

6-7 *Overhead valve keepers can be released with a simple spring compressor or, as explained in the text, shocked free with a wrench socket.*

An alternate disassembly technique is to position a large wrench socket over the collar and rap the socket sharply with a hammer. The impact will simultaneously compress the valve spring and dislodge split keepers. Assemble with a crow's-foot tool.

Valve springs are often—but not always—interchangeable between intake and exhaust sides. When springs differ, the heavier or double spring serves the exhaust valve. Some engines employ springs with closely wound damper coils that should be positioned on the stationary end of the spring (FIG. 6-8). In other words, the tightly wound coils are positioned farther away from the valve-actuating mechanism.

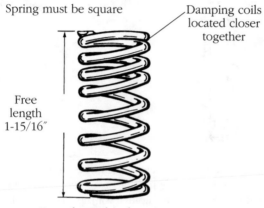

Spring must be square

Damping coils located closer together

Free length 1-15/16″

For valve-in-head engines

6-8 *Install variable-rate springs with the damper coils on the stationary ends, farthest from the actuating mechanism. Springs should be replaced as a part of every overhaul and especially on OHV engines.*

Weak valve springs can cause a hard-to-diagnose high-speed miss and, on OHV designs, can become detached with disastrous consequences. Nevertheless, original springs can reused if:

- The spring stands flat.
- The free-standing height meets the manufacturer's specification.
- There is no evidence of stress pitting or contact between adjacent coils.

Valve nomenclature is shown in FIG. 6-9. Inspect intake and exhaust valves for deep pitting, cracks and stem distortion. Leaks across the valve face cost compression, a condition signaled by hard starting, loss of power, and, on the intake side, by "pop back" through the carburetor.

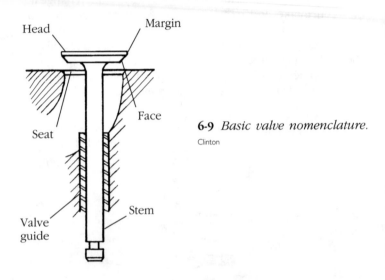

Head   Margin

Face

Seat

Stem

Valve guide

6-9 *Basic valve nomenclature.*
Clinton

Worn valve faces and seats should be turned over to a dealer or competent automotive machinist for servicing. The cost of the tools makes this work prohibitive for the casual mechanic.

Figure 6-10 shows a commercial valve grinder in use. While most small engine valves are cut at 45 degrees, Onan likes 44 degrees. Briggs & Stratton models have 30-degree intake valves and 45-degree exhausts. The moral is that valve work requires factory documentation for the particular make and model. At any rate, the valve face is cut at a single angle and should leave a margin (See FIG. 6-9) of about ¼₆₄ inch. Less will cause the valve to overheat and might send the engine into detonation.

The shop should also be able to handle seat refinishing (although automotive machinists do not always have the appropriate pilot). Normally, a high-speed grinder is used, but cast-iron, or integral, seats can be refurbished with a relatively inexpensive reamer of the type shown in FIG. 6-11. Valve seat angle and seat width are matters of specification, but the angle is always ½ to 1 degree larger or smaller than the valve face angle in order to provide an interference fit.

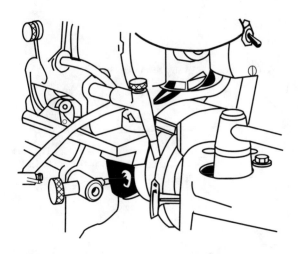

6-10 *Valve faces are resurfaced on a high-speed grinder, which can also be used to widen margins and dress stem ends.*

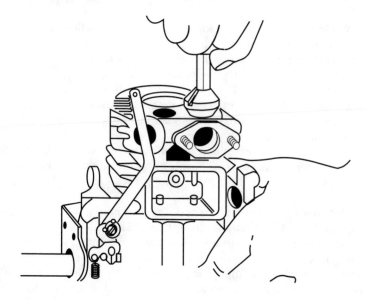

6-11 *Valve seats can be cut by hand or with a portable grinder.*

Seat width is controlled by entry and exit angles (as shown in FIG. 6-12). An overly narrow seat soon hammers out under valve impact and a seat that is too wide makes a poor seal.

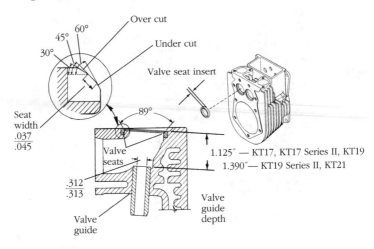

6-12 *Some idea of the crucial nature of valve geometry can be had from this Kohler-supplied illustration.*

While you always have some loss of control when work is farmed out, you should expect the machinist to accurately reproduce original valve dimensions. Provide the specifications; if he does not have them, and ask that grinding be kept to a minimum. "Buried" valves have poor flow characteristics and reduce available spring tension. Springs, however, can be shimmed at their stationary ends with hardened washers available from bearing supply houses.

Some manufacturers suggest that valves should be lightly lapped after machine refinishing. Few professional mechanics take the time to do this, but lapping does ensure a perfect seal.

Obtain a suction cup tool sized for small engine work (such as K-D Tools No. 501) and tin of Clover Leaf oil-based valve-grinding compound. Dab a small amount of compound on the valve face, insert the valve, and mount the suction cup tool. The cup might not find purchase on highly polished valve heads and some form of adhesive can be used. Rotate the tool between your palms, stopping when the compound degrades and no longer makes the characteristic hiss as the valve is worked. Raise the valve from its seat, spot a little more compound around the face, rotate the valve a quarter-turn, and repeat the operation. Stop when the valve face and seat take on a uniform metal finish. Wipe all traces of compound from the valve, seat the valve chamber, and flush with solvent. Compound that remains in the engine will continue its work on valve stems and guides.

At the risk of repetition, I should point out that valve lapping is a touchup operation that should not require more than 30 seconds per valve. Excessive lapping is counterproductive.

Valve guides

Integrity of the valve seat depends in large measure upon the condition of the valve guide that centers the valve (and reconditioning tool) on the seat. It is a waste of time to grind a valve that rides in a sloppy guide.

Guide-to-valve-stem clearance is measured at the top (valve head end) of the guide, and any figure of 0.0045 inch or more is excessive. Minimum clearance is quite small (on the order of 0.0015 inch.). Briggs & Stratton supplies their dealers with plug gauge sets to make these determinations. Experienced mechanics check valve guide wear as a function of how much the valve wobbles when fully open.

Most engines have some form of replaceable guide that is pressed into place. A few of the least expensive models run the intake and sometimes the exhaust valve directly against block metal, but there is always provision to retrofit guides. Many Briggs & Stratton guides are repaired by partial reaming and installation of a bushing in the upper guide area. And most manufacturers supply valves with oversized stems that can be installed after the original guides are reamed to fit.

If you establish that guides are worn, the first step is to obtain the necessary replacement parts and reamer. You should be able to find the reamer at a good tool supply house and the new guides, guide bushings, or oversized valves at a dealer (along with detailed installation instructions).

Replaceable guides are knocked out from the valve head end after measuring the distance between the top of the guide and the valve seat insert. Some side valve guides must be broken before extraction from the valve chamber; others can be withdrawn in one piece. Clean all parts in solvent and—working from the valve head end—drive the replacement guides home to the depth of the originals. Valve guide drivers that center on the guide inside diameter can be purchased from specialty tool houses. Use of the appropriate driver can eliminate the need to finish-ream the guide to 0.0015–0.002 inch larger than the valve stem. Otherwise, the guide must be reamed.

Figure 6-13 shows how Briggs & Stratton valve guide bushings are superimposed upon the original guides. This job requires special tools and should be farmed out to a dealer.

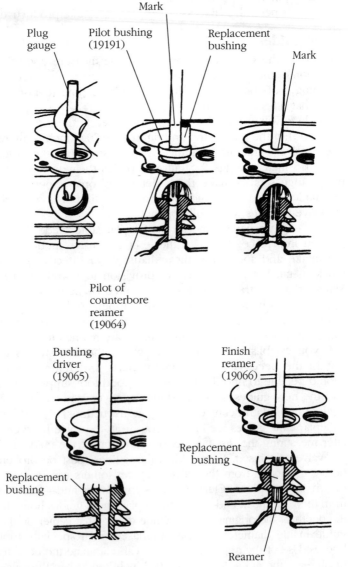

6-13 *Installation of valve guide bushings in Briggs & Stratton engines is a job best left to a dealer.*

Figures 6-14 and 6-15 depict another approach to guide service. In this example, guides are mounted in an aluminum head. The head is immersed in oil and heated to 375–400 degrees F—a process that should be done outdoors—and the old guides are pressed out. Then the head is brought back up to temperature and new guides, which have been chilled, are pressed in. Because installation depends upon thermal expansion and contraction, little violence is done to the guides. Finish reaming should not be necessary.

Note: Valves and valve seats must be ground after guide installation.

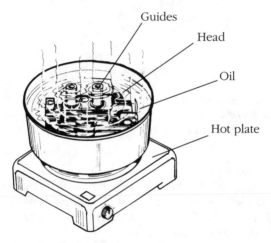

Guides

Head

Oil

Hot plate

Heat until oil begins to smoke

6-14 *Aluminum OHV heads do not take kindly to brute-force methods of valve-guide extraction and installation. A better technique is to heat the head in oil (supporting off the bottom of the container).*

Valve seats

Most engines feature replaceable valve seats that must be renewed in event of looseness, cracking, or deep pitting (a seat insert is shown in FIG. 6-12). Briggs & Stratton cast iron block models employ a replaceable exhaust valve seat, but the intake valve runs directly on the block. Replacement intake valve seat inserts can be installed with proper tools as described.

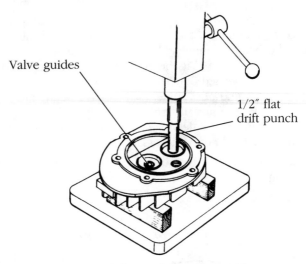

Valve guides

1/2″ flat
drift punch

Center drift punch on valve guide

6-15 *If the temperature differential between the guide and the head is great enough, the guide would drop into place. As a practical matter, an arbor press is needed.*

Valve seat replacement is a relatively unusual service operation and you should obtain parts before embarking upon such work. Drive the old seat out with a long punch inserted through the port. If that is not possible, use a purchased or homemade removal tool (FIG. 6-16).

Most replacement valve seats—particularly those intended for installation in cast-iron blocks—have the same outside diameter as the original seat, eliminating the need for a special reamer, but limiting the repairability of the block when the seat has loosened and wallowed its recess. The wear limit for Briggs & Stratton engines is 0.005 inch clearance between the seat outside diameter and block inside diameter (FIG. 6-17). The alternative approach was typified by Clinton in their GEM series engines. Replacement seats were 0.040 inch oversized and a reamer, piloted on the valved guide, was used to enlarge the recess.

As mentioned earlier, intake valve seats on Briggs & Stratton cast iron engines must be reamed to accept an insert. All aluminum block engines employ hard inserts that must be renewed when cracked or loose.

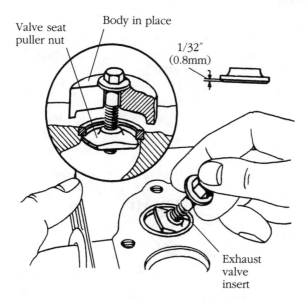

Valve seat
puller nut

Body in place

1/32″
(0.8mm)

Exhaust
valve
insert

6-16 *When port geometry makes it impossible to drive the seats out with a punch, a bushing puller (available from auto supply stores) or Briggs & Stratton PN 19138 can be used.*

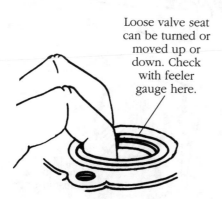

Loose valve seat
can be turned or
moved up or
down. Check
with feeler
gauge here.

6-17 *The only way to repair loose valve seats is to ream the boss to accept an oversized replacement seat. Adhesives or staking does not work.*

The replacement seat and recess must be dry and spotlessly clean. Chill the seat and, working quickly, press it into place. Use a factory driver if available. Otherwise, use a scrap valve. It is good practice—especially on aluminum block engines—

to stake the newly installed seat. Begin at three points 120 degrees apart and complete the job at close intervals around the seat circumference (FIG. 6-18).

The associated valve should be replaced or, if still serviceable, reground and lapped to the seat.

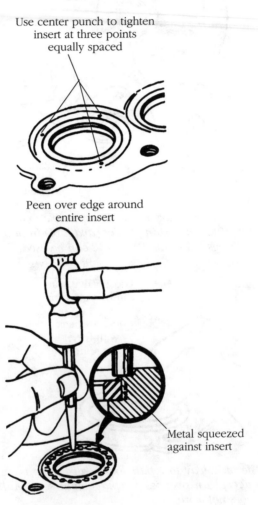

Use center punch to tighten insert at three points equally spaced

Peen over edge around entire insert

Metal squeezed against insert

6-18 *Some mechanics peen seat inserts in aluminum castings, although the value of the practice can be questioned. A better fix is to cut the boss so that the top of the seat is recessed about 0.030 in. below the surrounding metal. Then, using a hammer and dolly, roll the metal over the seat outside diameter.*

Valve lash adjustment

Metal lost through grinding and lapping operations and metal gained through seat or valve replacement must be compensated for by valve lash adjustment. Side-valve engines are generally fitted with nonadjustable tappets, and valve lash increases are made by grinding the valve stems. Install the valve without spring, turn the crankshaft until the tappet fully retracts, and measure the clearance between the tappet and valve stem with a feeler gauge (FIG. 6-19). Grind the valve stem as necessary to establish factory-specified clearance (on the order of 0.008 inch intake and 0.010 inch exhaust). Work slowly, frequently rechecking the lash, and take care that the stem remains dead flat. If too much metal is lost, the lash will be excessive and valve timing will retard for some loss of power. Correct by regrinding and lapping the valve face.

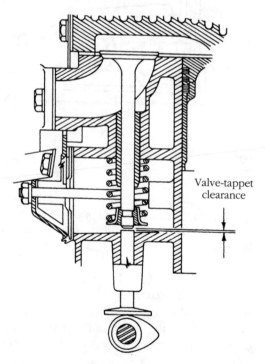

Valve-tappet clearance

6-19 *On side-valve engines, valve lash is measured between the end of the valve stem and the tappet, with the tappet on the cam circle. Nonadjustable tappets are the norm for SV engines. Adjustment is possible by grinding the seat or valve stem end.*

Some of the better side-valve engines and all overhead valve types have adjustable tappets. Valve lash for OHV engines is defined as clearance between the rocker arm and valve stem (FIG. 6-20). Turn the crankshaft until the associated tappet is on the heel of its cam lobe, loosen the lock nut, and turn the adjustment nut to achieve specified clearance. Tighten the lock nut and recheck.

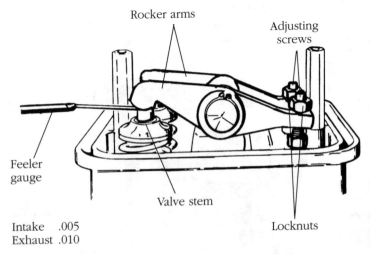

Turn adjusting screw clockwise to decrease lash, counterclockwise to increase lash.

Rocker arms

Adjusting screws

Feeler gauge

Valve stem

Intake   .005
Exhaust .010

Locknuts

6-20 *OHV lash is measured between the end of the valve stem and the rocker arm. Because of the remoteness of the camshaft, manufacturers of OHV engines always include a provision for lash adjustment.*

Valve gear modification

It is sometimes possible to upgrade standard-duty engines to heavy-duty status by parts substitution. TABLE 6-2 gives the interchange parts numbers for Briggs & Stratton engines.

Breather

Four-cycle breather engines incorporate some form of breather assembly (FIG. 6-21). The breather consists of a check valve, bleed port, and oil trap. The check valve opens on the piston

Table 6-2. Briggs & Stratton Stellite Valve and Torocap Conversion.

	Stellite Valve	Rotocap only conversion			
		Spring	Rotocap	Retainer	Pin
⊿min					
◖00, 80000, 92000, ◖00	260443	26826	292259	230127	230126
◖000, 13000	260860	26826	292259	230127	230126
◖000, 170000, 190000, ◖000, 250000	390420	26828	292260	93630	
st iron					
19, 190000, 20000	26735	26828	292260	68283	
230000	261207	26828	292260	68283	
◖000, 300000, 320000	261207	26828	292260	68283	(Stellite Std.)

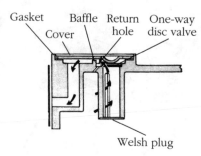

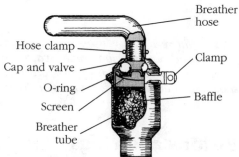

6-21 *A small engine breather assembly includes a disk (A) or ball-type (B) check valve, a baffle to separate liquid oil from vapor, a provision for liquid oil to drain back into the crankcase, and a vapor vent. The latter usually feeds into the carburetor.*

downstroke—when crankcase pressures are highest—to allow air to be expelled from the crankcase. It remains closed for the rest of the stroke, sealing the crankcase and keeping its pressure at some value below atmospheric.

Partial vacuum tends to reduce oil seepage at gaskets and crankshaft seals. However, the breather bleed port remains open during the whole cycle. Consequently, some air enters the crankcase to be circulated and expelled—together with combustion gases—on the next downstroke. The oil trap prevents escape of lube oil out of the breather tube. Exhaust can be vented to the atmosphere or to the carburetor intake.

Some two-cycle engines use a functionally similar device for an entirely different purpose. The reed assembly (FIG. 6-22) acts as a check valve to contain the air-fuel mixture in the crankcase.

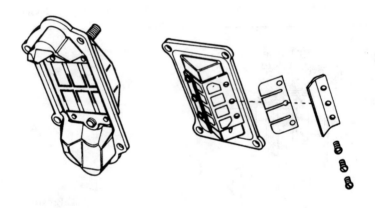

6-22 *A flat reed valve assembly is one of the marks of a low-performance engine; the greater valve area of the pyramidal assembly on the right promotes crankcase filling. The unit shown includes a stop plate to limit petal travel.*

Pistons and rings

Single-cylinder utility and industrial powerplants typically have their cylinders cast in one piece with the block. Horizontal crankshaft engines give access to the crankshaft by means of a side cover and/or removable oil sump. Vertical-crankshaft crankcases are split just under the cylinder. The PTO-side casting, known as the flange, supports the PTO main bearing and one end of the camshaft.

Disengage the flange or side cover as described in the caption to FIG. 6-23. The split connecting rods commonly found on these engines are secured to the crankpin by bolts and studs. To avoid catastrophic assembly error, reference both the rod cap-to-rod-shank orientation and the orientation of the rod and piston assembly to some prominent crankcase feature. These matters are discussed in the "Connecting Rod" section. Using a hammer handle or wood dowel, drive the piston and attached rod shank out of the top of the bore (FIG. 6-24).

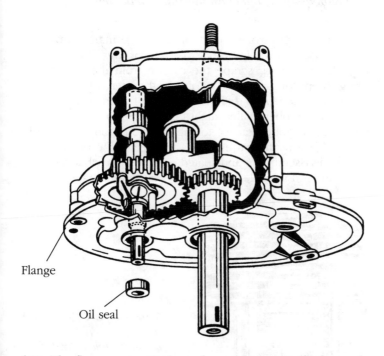

Flange

Oil seal

6-23 *The flange on vertical-crank engines locates the lower, or PTO, main bearing. Before extraction, carefully remove rust and tool marks on the crankshaft with emery cloth and a file. Lubricate the crankshaft and, with the engine mounted vertically, remove the flange holddown bolts. Separate the castings with a rubber mallet. The camshaft should remain engaged with the flywheel so that timing mark orientation can be verified. In no case should you attempt to pry the flange off.*

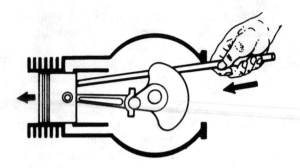

6-24 *Once the rod cap is detached, use a wood dowel to drive the piston assembly out of an integral cylinder barrel.*

For better control of port dimensions, many two-cycle engines use removable cylinder barrels, or *jugs*. With the barrel still assembled, scrape the carbon from the piston top. Bring the piston down to bottom dead center (BDC), remove the barrel fasteners, and lift the barrel off the crankcase. A few raps with a rubber mallet might be required to break the gasket seal. It is good practice to support the piston as shown in FIG. 6-25 before the barrel comes free. Raise the piston; if no further disassembly is anticipated, stuff the area between the block and rod with clean rags to prevent dirt infiltration.

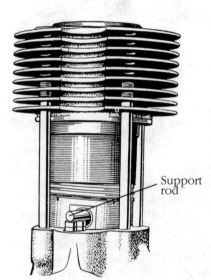

Support rod

6-25 *Support the piston when lifting a demountable cylinder barrel.*

Figure 6-26 shows the crankcase architecture of the typical single-cylinder two-stroke engine. The crankcase parting line passes through the center of the cylinder bore. Each half of the crankcase carries a main bearing, nearly always secured by a press-fit and difficult to disengage without the proper tool. In contrast, multicylinder two-cycles and some imported single-cylinder two- and four-cycle units split horizontally along the centerline of the crankshaft. These engines open like a book.

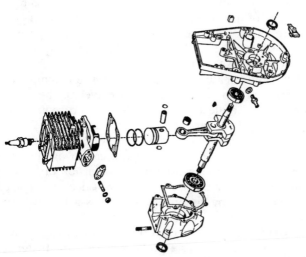

6-26 *Splitting the crankcase on two-cycle engines with detachable cylinder barrels can pose difficulties. The resistance of main bearing press fits, sealant (applied to the crankcase parting line), and locating pins must be overcome before the case halves can be separated. Usually one or both main bearing bosses have been drilled and tapped to accommodate a puller. This tool, which is similar to those used to remove harmonic balancers from automobile crankshafts, can be purchased from the engine maker or fabricated. If bearing bosses are not drilled, the cases can be gently warmed with a propane torch and pried apart with a hammer handle inserted into the cylinder-barrel cavity. Exercise extreme care to avoid warping the fragile castings.*

A few high-performance two-cycle engines (i.e., engines with detachable cylinder barrels) use one-piece connecting rods. See the "Connecting Rod" section below for details.

Inspection

Bright rings, uniformly polished and with no vestige remaining of tool marks, are simply worn out. Stuck rings, frozen into their grooves, indicate poor maintenance, extreme service, excessive combustion temperature, and, as a consequence, loss of ring spring tension. Broken rings have several causes, including inept installation, detonation impact, and worn grooves that allow the rings to twist during stroke reversals.

Chronic detonation can also affect the piston, nibbling at the crown as if mice were at work. The damage usually starts at one edge of the crown—adjacent to the area in the chamber where the fuel charge is slowest to ignite—and progresses toward the center. Check for poor fuel antiknock quality, lean carburetion, excessive ignition advance, and any other condition that would lead to elevated combustion temperature. Loads imposed too early in the rpm curve can be a factor, because large throttle openings at low speed reduce turbulence and slow flame propagation.

Preignition is rare but obvious when seen from the vantage point of the piston. The center of the crown overheats and can dent or burn through from the combination of high temperatures and premature gas expansion. Check the combustion chamber for any abnormality—such as a hangnail spark plug thread, a piece of partially detached carbon, or a knife-edged exhaust valve—that could produce a constant source of ignition. Grinding two-cycle ports oversized for better flow and increased power sometimes has the same result because the bridge between exhaust ports is narrowed and can become incandescent.

Examine the piston skirt for wear. Typically, rubbing contact occurs at two points at right angles to the wrist pin centerline and gradually expands to the whole length of the skirt. Figure 6-27 shows abnormal wear patterns produced by bent and twisted connecting rods. Forces that rocked the piston to make these patterns can also drive the wrist pin past its locks and into contact with the cylinder wall.

Deep scratches suggest a cylinder bore problem. Light abrasions, giving the piston a matte finish, point to air filter leaks. Once sand has been ingested, all bearings become contaminated and the engine should either be scrapped or completely rebuilt.

Utility and industrial engines are set up fairly tight, with piston-to-bore clearances between 0.0015 and 0.002 inch. How

A

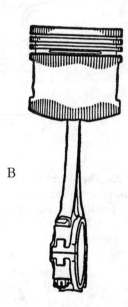

6-27 *A bent conn rod tilts the piston to create an hourglass-shaped wear pattern, shown by the shaded lines in drawing A. A twisted rod rocks the piston, concentrating wear on the upper and lower edges of the skirt, illustrated in drawing B.*

B

much wear is tolerable is, in part, a subjective judgment involving tradeoff between immediate cost and anticipated life to the next overhaul. Most factories put the wear limit at 0.005 or 0.006 inch, but small-bore high-rpm engines are happier if piston clearance does not exceed 0.004 inch.

Pistons usually taper toward their crowns at a rate of about 0.00125 of an inch per inch of height, which allows the hottest part of the piston room to expand. In addition, four-cycle pistons are cam-ground so that thrust faces are on the long axis. The piston remains centered on the bore when cold and gradually expands to a full circle as the engine warms to operating temperature. Two-cycle pistons are sometimes round, rather than oval, to control crankcase leakage during startup.

All measurements are made across the thrust faces, at right angles to the wrist pin. Manufacturers specify a distance above the base of the skirt and just under the wrist pin. Clinton backstops skirt diameter with a measurement across the second ring land (FIG. 6-28).

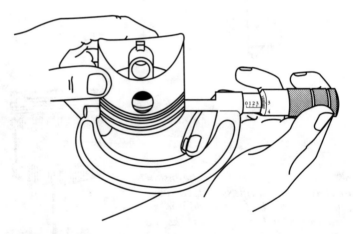

6-28 *The primary piston measurement is made across thrust faces at some specified distance below the piston pin in recognition of piston taper. Straight-cut (untapered) pistons can be checked at some point near the upper ring land, where wear is greatest.*

The final piston check is to determine ring groove width. Remove the rings and scrap all traces of carbon from the grooves, opening oil drain holes in the lowest groove. You might want to use a special groove cleaning tool, available at auto parts stores, or a broken piston ring mounted in a file handle.

Warning: Piston rings—especially used rings—are razor sharp.

Using a new ring, measure side clearance on both compression ring grooves (FIG. 6-29). Excessive side clearance (as defined by the manufacturer) allows the ring to twist during stroke reversals (FIG. 6-30). This condition defects ring sealing geometry and eventually causes breakage. While it is theoretically possible to recut the grooves overly wide and restore clearance with spacers, the best option is to replace the piston.

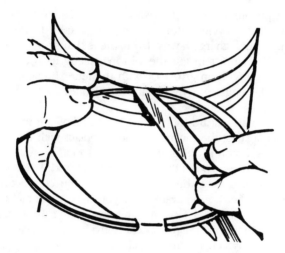

6-29 *Determine ring side clearance with a new ring as reference. The upper side of No. 1 groove (shown) takes the worst beating.* Ona

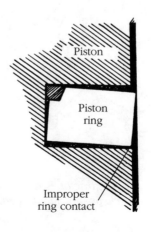

6-30 *A major cause of ring breakage (along with improper installation and detonation) is the twist created by worn ring grooves.*

Piston

Piston ring

Cylinder wall

Improper ring contact

Piston pin

Four-cycle wrist pin bearing wear is relatively uncommon because thrust reverses every second revolution. In contrast, two-cycle pins are subject to an almost constant downward force that tends to squeeze out what lubrication is present. In either case, the bearing is considered acceptable if it has no perceptible hand up-and-down play and if the piston pivots of its own weight.

Most pistons incorporate a small offset relative to their pins and some two-cycle piston crowns are shaped to deflect the incoming fuel charge away from the exhaust ports. Consequently, you must install the piston exactly as found. Some are stamped with an arrow or with the letter F (signifying the front of the engine). Others can be oriented by the manufacturer's logo.

Remove and discard the circlip. New circlips are inexpensive insurance against the pin moving into contact with the cylinder bore. If the piston is out of the engine, support it on a wood V-block and drive or press the pin clear of the rod. Do not gouge the pin bore during this process. When the connecting rod remains attached to the crankshaft, you have to extract the pin with the tool (FIG. 6-31) or by carefully heating the piston. Do not heat with an open flame. Besides inviting a crankcase explosion, this approach is almost guaranteed to distort the piston. Instead, wrap the piston with a rag soaked in hot oil or—less messily—heat the crown with an electric hot plate.

6-31 *A piston pin extractor is a useful tool that can be ordered through motorcycle or snowmobile dealers. A Kohler tool is shown.*

Installation is essentially the reverse process, except that piston pin and pin bores must be well lubricated. Make certain that the new circlips seat in their grooves.

Piston rings

Four-cycle pistons employ three distinct ring types. Counting from the bottom, there is the oil control ring (cast in one piece or made up of several steel segments), the scraper, and the top compression ring. Four-ring pistons employ a backup compression ring. Some manufacturers offer the option of engineered replacement ring sets. These sets include expanders behind the compression, the scraper, and sometimes the oil control ring to increase ring tension. This expedient permits better conformity with worn bores, but costs something on the order of 2 percent loss of power and additional fuel consumption.

Two-cycle engines are fitted with two identical compression rings that do, however, have definite upper and lower sides. Installing the rings upside down will cost compression and power. Pegs can be used to secure ring ends to prevent rotation and possible hangup in cylinder ports.

Determine the end gap of each ring as verification that the correct diameter rings are installed and as a final check on cylinder-bore dimensions. Using the piston crown as a pilot to hold the ring square, insert the ring about midway into the cylinder (FIG. 6-32). Measure the end gap with a feeler gauge.

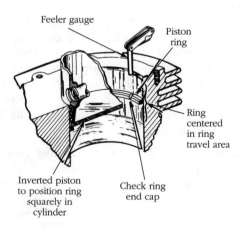

6-32 *Using the piston as a pilot, insert each replacement ring about halfway into the bore and measure its end clearance.*

Specifications vary, but most manufacturers call for about 0.0015 inch of ring gap per inch of cylinder diameter. Too large a gap wastes compression and indicates an undersized ring or oversized cylinder. Too narrow a gap can allow the ring ends to abut under thermal expansion, resulting in rapid cylinder wear and premature ring failure. Correct this by filing the ends; keep them flat and square.

Lay out the rings in the order of installation. Make certain that you have identified each ring and each ring's upper side, which can be marked as such (FIG. 6-33). Using the proper tool, install the oil ring from the top of the piston, spreading the ring only wide enough to clear the piston diameter. Repeat this operation for the remaining rings (FIG. 6-34). Verify that the rings ride in their grooves and, where applicable, that ring ends straddle locating pegs.

Rotate floating rings to stagger the ring gaps some 120 degrees so that there will be no clear channel for compression leakage. On Tecumseh engines with relieved valves (shades of Ford hot rod days!), it is important to position the ring ends away from the bore undercut (FIG. 6-35).

Installation

Integral barrel   Turn the crankshaft to the bottom dead center position and press short pieces of fuel hose over the rod studs. Lubricate the cylinder bore, crankpin, rod bearing, pin, and piston (flooding the rings with motor oil). Without upsetting the ring gap stagger, install a compressor tool of the type shown in FIG. 6-36 over the piston. Tighten the band only enough to squeeze the rings flush with piston diameter.

Position the piston and attached upper rod as originally found in the engine, and carefully tap the piston out of the compressor. Do not force the issue. If the piston binds, a ring has escaped the tool or there is interference between the rod shank and the crankshaft. Read the "Connecting Rod" section before installing the rod cap.

Detachable barrel   Lubricate the cylinder bore, piston pin, and piston ring areas. Support the piston on the crankcase with a rod (as shown in FIG. 6-25), or with a wooden fork (FIGS. 6-37 and 6-38). Some barrels are beveled and can be slipped over the rings without difficulty. Others are straight-cut, and require use of a clamp-type tool shown in FIG. 6-37.

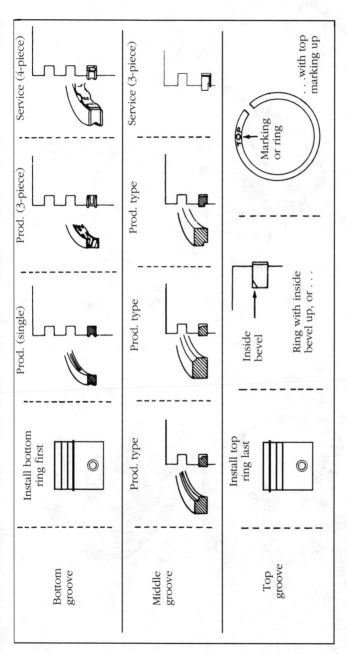

6-33 *Kohler ring sequence and orientation is typical of four-cycle engines.*

6-34 *Installing a compression ring on an Onan piston with the aid of a ring expander.*

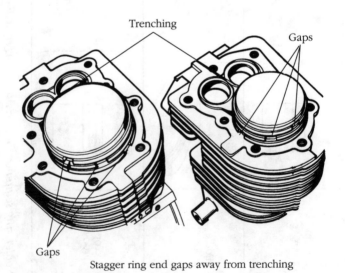

Stagger ring end gaps away from trenching

6-35 *Rings for Tecumseh engines with relieved, or trenched, valves must be installed with their ends turned away from the bore undercut.*

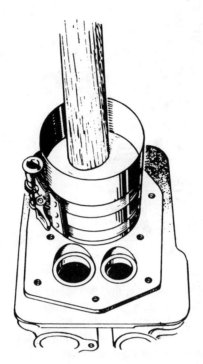

6-36 *A ring compressor sized for small engines is used when the piston is installed from the top of the bore. Using a hammer handle, gently tap the piston home.*

6-37 *A homemade ring clamp and a wooden rod-holding fixture make ring installation easier for many engines with detachable cylinder barrels.*

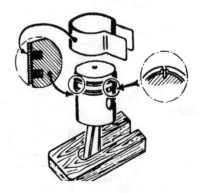

Cylinder bores

Pistons for cast-iron engines run directly on block metal without the intermediary of a cylinder liner. Iron provides a fairly

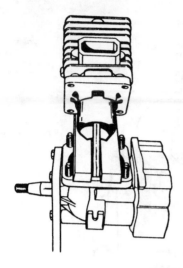

6-38 *Some detachable barrels feature chamfered bores to simplify ring installation. Note the use of a rod-holding fixture.*

durable wearing surface and flows readily enough to mend small scratches. Aluminum block engines are protected from ring scuff with a layer of chrome applied to the base metal or, as is more often the case, with an iron liner. A few linered bores are chromed for extreme wear and corrosion resistance.

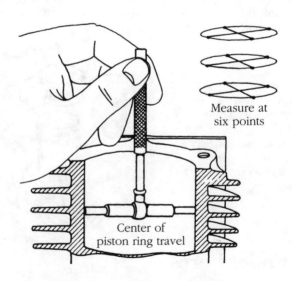

Measure at six points

Center of piston ring travel

6-39 *Measure the bore at six points to determine oversize, out-of-round, and taper.*

Inspect for deep scratches, aluminum splatter from piston melt, and for chrome separation. The plating is most vulnerable at the top of the bore and around the exhaust ports on two-cycle engines where thermal expansion is greatest. Rechroming the bore is impractical and any evidence of peel means that the block should be scrapped.

Maximum wear occurs near the upper limit of ring travel, where heat is greatest, lubrication is minimal, and corrosives are most concentrated (FIG. 6-39). On unchromed bores, wear results in a ridge at the upper limit of ring travel (FIG. 6-40). The amount of ridge is a rough guide to bore wear and can be significant enough to hinder piston extraction. In any event, the ridge must be removed before new rings are installed.

Possible ridge
buildup area

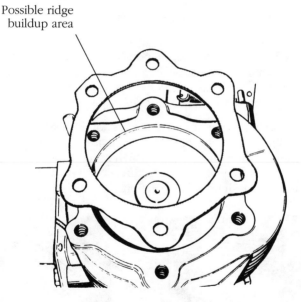

6-40 *The ridge that develops at the top of cast-iron bores must be removed or reduced before new rings are installed.*

Figure 6-41 shows a ridge reamer in use. Adjust cutter tension with the upper nut and rotate the tool clockwise. Insufficient tension dulls the cutter, while too much tension produces chatter and can fracture the carbide cutting edge. Lubricate the tool frequently and stop when the ridge is partly obliterated. No used cylinder is a perfect circle, and some evidence of ridge will remain at the long axis.

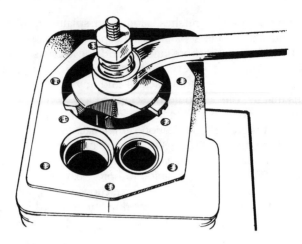

6-41 *Reaming the ridge on an Onan engine.*

Cast-iron bores develop a polished glaze that must be re-moved for new rings to seat. A spring-loaded home that auto-matically conforms to cylinder diameter is preferred for "glaze busting," although patience and sandpaper will do much.

1. Chuck up the hone in a drill press or a ½-inch portable drill motor. A smaller drill will provide sufficient power, but most turn at excessively high speeds. Anything more than 400 rpm will produce "threaded" surface (FIG. 6-42) and defeat the whole operation.

Avoid this finish

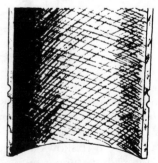

Produce cross hatch scratches for fast ring seating

6-42 *A honed cylinder should consist of thousands of diamond-shaped points that retain oil and wear in quickly.* Onan

2. Fit the tool with a 280-grit stone and lubricate as the tool manufacturer suggests.
3. Cycle the hone about 70 times a minute with ¾ inch or so of stone protruding from each end of the cylinder at extremes of travel (FIG. 6-43).
4. Stop when the cylinder bore is uniformly scored.
5. Scrub the bore with brush and detergent to remove every trace of abrasive. Cleanup cannot be accomplished with solvent.

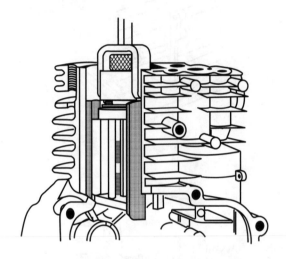

6-43 *Cutaway view of Clinton block shows a precision hone at its lower travel limit. Note the extension of the stones beyond the bore.*

Cylinder oversizing is a more serious matter and requires a precision hone. If a lathe is used, the last few thousandths of the cut must be honed to remove tool marks. Most mechanics find it easier to use a hone for the whole operation:

1. Adjust drill press for 300 to 400 rpm spindle speed.
2. Select a coarse (80 grit) stone for initial cuts.
3. Position the work piece on the tool table. The bore must be vertical and, at the same time, free to move laterally. Figure 6-44 shows an arrangement using shims for vertical alignment. The table can be oiled to further aid centering.
4. Set the press stops to allow the stones to extend ¾ inch beyond both ends of the bore.

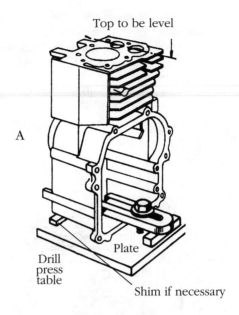

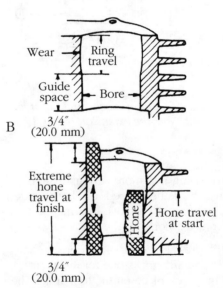

6-44 *Mount the block or barrel loosely on the drill press table to aid alignment. Vertical-shaft blocks must be shimmed to level the fire deck with the table (A). When resizing, begin at the lower and least worn portion of the bore (B).*

5. Adjust the hone to conform with the lower bore diameter (which will be smaller than upper bore). Contact should be positive, but not so firm as to prevent turning the tool by hand.

6. Lubricate the stones as the manufacturer suggests. Petroleum solvents can dissolve the binder and cause rapid stone wear.

7. Start the press and keep the hone moving about 70 strokes a minute. As the lower cylinder enlarges, adjust the hone for greater reach. Eventually the whole length of the cylinder will be traversed on each stroke.

8. Keep a close eye on how much metal is removed. The bore can tend to "bell mouth" at the ends.

9. Clean the felts and replenish lubricant at frequent intervals.

10. When the bore is straight and within 0.002 inch of final size, stop and change to a 280 finishing stone.

11. Make the final cut; stop frequently for a dimension check. Verify running clearance against the replacement piston. The actual diameter may vary 0.0005 inch from nominal diameter.

12. Scrub the bore with a brush, hot water, and detergent. Wipe dry with paper shop towels. Scrub until the towels are no longer stained with abrasive; oil immediately.

Connecting rods

Aluminum is the material of choice for four-cycle connecting rods that, almost always, are split at the big end. Utility and light industrial engines do not have replaceable bearings; crank and piston pins run against the rod itself. Figure 6-45 shows the standard pattern that, in this case, incorporates an oil slinger below the cap for splash lubrication.

Better-quality engines use precision bearing inserts at the big end in conjunction with a bushing at the small end (FIG. 6-46). Undersized inserts (0.010 and 0.020 inch for American-made engines) allow the crank to be reground.

Traditionally, two-cycle connecting rods were steel forgings with needle bearings at both ends. However, light and moderate-output power plants, including some outboard motors, currently use aluminum, which can itself act as the bearing. Plain bearing engines are adequate for light duty in applications if a 1

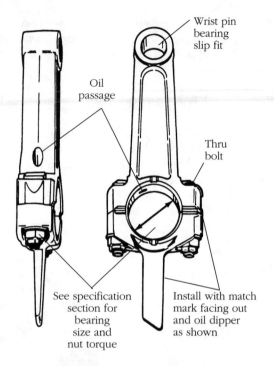

Wrist pin
bearing
slip fit

Oil
passage

Thru
bolt

See specification
section for
bearing
size and
nut torque

Install with match
mark facing out
and oil dipper
as shown

6-45 *The typical white-metal four-cycle connecting rod runs on integral bearings. There oughta be a law!*

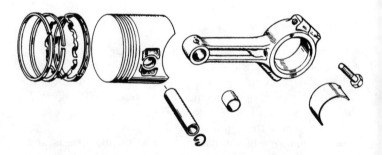

6-46 *Onan conn rod features a bushed small end and precision inserts at the big end.*

to 24 oil to gasoline ratio can be tolerated. As a point of comparison, an automobile engine is considered worn out if oil consumption is 1 to 400.

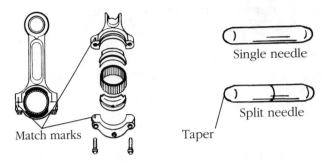

Single needle

Split needle

Match marks   Taper

6-47 *Two-cycle rod with needle bearings at the lower end for reduced lubrication requirements and durability. Needle bearings also increase the rev limit because two-cycle engines do not deliver the flood of oil demanded by plain bearings.*

Figure 6-47 shows a connecting rod for a two-cycle industrial power plant. The aluminum rod rides on a bushing at its upper end and on single-row or double-row needles at the crank end. Note the use of replaceable races.

Catastrophic rod failure almost always originates at the big end. How it happens is, in part, a function of big end bearing type. Plain bearings skate on a pressurized wedge of oil that appears soon after startup. Once up to speed, the bearing should, in a sense, hydroplane and make no direct contact with its journal.

Insufficient clearance between the crankpin and bearing prevents the oil wedge from forming; excessive clearance allows the wedge to leak faster than it can be formed. In either case, the result is metal-to-metal contact, fusion, and a broken connecting rod.

Needle bearings make rolling contact against their races without the cushion of an oil wedge. Consequently any imperfection—fatigue flaking, rust pitting, skid marks—means bearing seizure and rod failure.

Split big ends occasionally crumple into bite-sized chunks as a result of insufficient rod-bolt torque. Proper torque might not have been applied during assembly, or rod locks might have given way, allowing the bolts to shake loose. This is why the manufacturer's torque specifications must be followed to the letter and why new lock washers or locknuts must be installed whenever a connecting rod is disassembled. Bend-over tab locks usually carry a spare tab that can be employed during the first overhaul. Once the tab is engaged, it should not be straightened and reused.

Orientation

Correct orientation is vital and, counting the piston, has three aspects:

- Piston-to-rod. The piston pin can be offset relative to the piston centerline and two-cycle pistons can be asymmetrical.
- Rod assembly-to-engine. Some connecting rods are drilled for oil and vapor transfer; others are configured so that reverse installation locks the crankshaft.
- Cap-to-rod. In order to maintain the necessary precision, most engine makers assemble the rod and cap and ream to size.

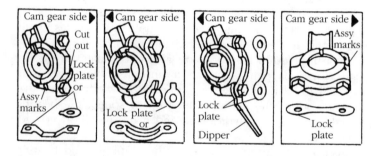

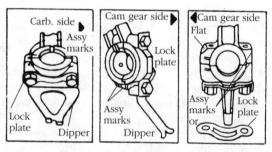

6-48 *Briggs & Stratton rod-to-engine and cap-to-rod orientation. Another example of embossed rod and cap index marks can be seen in "A" of 6-47. McCulloch engines employ steel rods that are fractured after machining. When assembled correctly, the parting line becomes almost invisible.*

Stamped or embossed marks identify cap orientation (FIG. 6-48). Failure to assemble the cap correctly results in early and catastrophic failure.

Inspection

The pin should be loose enough on the rod to allow the piston to pivot of its own weight at room temperature. Pin-to-piston fits are tighter, but loosen when the piston reaches operating temperature.

Small-end bushings should be replaced and reamed to size when worn or loose on the rod eye. Needle bearings, often used on the small end of two-cycle rods, suffer from insufficient lubrication because of the constant downward pressure developed by these engines.

The big end bearing is the most critical friction surface in the engine and never more so than when in the form of needles or rollers. Red rust stains or any other visible imperfection on bearings or contact surfaces mean that the whole assembly should be replaced. That assembly includes the crankshaft and rod. (Briggs & Stratton makes this mandatory by supplying needle-bearing cranks only as built-up assembles, with connecting rod and bearings mounted.)

Wear on the crankpin can be determined by comparing pin diameter with the manufacturer's specification. Wear on needle bearings cannot be measured directly. The best that you can do is assemble the bearing dry (with rod match marks aligned) and make a judgment based on the amount of wobble.

The crankpin is the critical element for plain bearing rod assemblies (the connecting rod or insert bearings can be replaced at nominal cost). Measure the crankpin at several places along its length and diameter with a micrometer. A comparison of average diameter with factory specifications will show the amount of wear; taper and out-of-round should within 0.001 inch.

Once the crankpin diameter is known, the rod is assembled, match marks—as always—together, and its diameter is measured.

The difference between average crankpin and rod bearing diameter equals running clearance, which is subject to specification, but which should fall between 0.001 inch on a new and somewhat tight assembly to, say, about 0.004 inch (which is pushing things a bit). Do not attempt to restore bearing clearance by filing the ends of the rod cap. This expedient does not work for very long.

Another way to establish bearing clearance and crankpin asymmetry is to use plastic-gauge wire (available from auto parts jobbers). The soft, plastic wire is precisely dimensioned and flattens as the bearing cap is installed. Wire width converts to bearing clearance via a scale on the package. Follow this procedure:

1. Turn the crankshaft and assembled rod to bottom dead center.
2. Remove the rod cap.
3. Wipe off all oil on the rod cap and exposed crankpin.
4. Tear off a piece of gauge wire and lay it along the full length of the crankpin (A of FIG. 6-49).

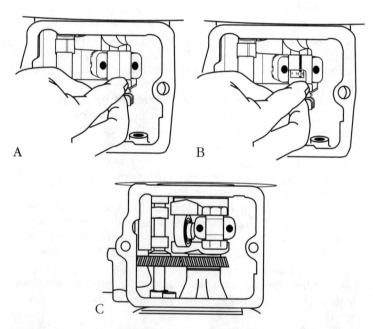

6-49 *Lay a piece of plastic gauge wire along the length of the crankpin (A). Install the cap and—without moving the crankshaft—torque the rod nuts to factory specification. Lift the cap off and measure the flattened wire against the scale on the package (B). Repeat the operation, positioning the wire at two points on the crankpin circumference to detect taper (C).*

5. Install the rod cap, oriented correctly, and pull it down evenly to factory torque specifications.

Caution: Do not rotate the crankshaft during this procedure.

6. Remove the cap and measure the width of the gauge wire against the scale printed on the envelope (B of FIG. 6-49). Average width corresponds to bearing clearance; variations in width from one end of the crankpin to the other show taper.

7. Repeat the process, as shown in C of FIG. 6-49. This is a cross-check on taper and indicates out-of-round.

Even if you prefer to make the initial determination with precision gauges that seem to show taper and out-of-round more positively than wire, the final check on the installed bearing clearance should be made with plastic. This positive measurement has very little room for errors of interpretation.

Assembly

Coat upper and lower bearing surfaces thoroughly and liberally with clean motor oil. Failure to do this can ruin a bearing on initial startup. The insert type of big-end bearings sometimes has an oil hole that defines the upper end. Otherwise, bearing inserts interchange between rod cap and shank. Uncaged needle bearings can be fixed around the periphery of the crankpin with grease or, following the old practice, with beeswax. Protect the crankpin during piston installation with short lengths of fuel line over the rod bolts.

The full-circle rod, used on Tecumseh TVS and TVXL840 engines, installs with the flange toward the PTO side of the engine (FIG. 6-50A). Replacement bearings are packed in a viscous grease that should hold them in place while the rod is maneuvered over the crankpin (FIG. 6-50B).

Check the piston-to-block, piston-to-rod, and rod-cap orientation one final time. Turn the crank down to BDC and, using your fingers, guide the rod assembly home. Install the correctly oriented cap and new rod locks, and run the bolts down to specified torque, keeping the cap square during the process.

Pull the engine over by hand for several revolutions to detect possible binds. The rod should move easily from side to side along the crankpin. Most manufacturers do not provide a side play specification, but the rod is comfortable with several thousandths of an inch of axial freedom.

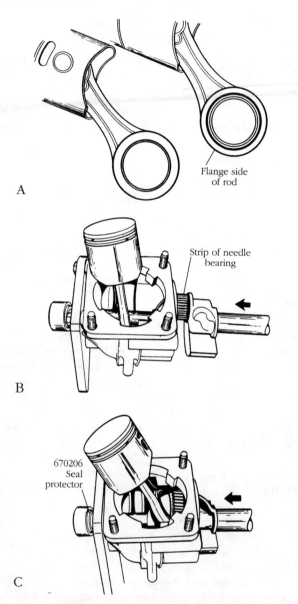

Flange side
of rod

A

Strip of needle
bearing

B

670206
Seal
protector

C

6-50 *TVS and TVXL840 rods present a special case. Rod installs with flange toward the PTO side of the engine (A). Self-adhesive bearing strip (28 needles) is applied to the crankpin (B), and rod gingerly positioned over the crankpin and bearings (C). Note the factory-supplied seal protector, which is one of two that are needed for this job.*

Crankshafts

It is always good practice to align timing marks before four-cycle engines are disassembled. Crankshaft and camshaft timing marks index at top dead center (TDC) on the compression stroke. Secondary marks on rotating balance or accessory-drive shafts are indexed to the crank or cam after primary alignment is made.

Occasionally, timing marks wear away and you must time the engine from the "rock" position. Rotate the crankshaft to bring No. 1 piston to top dead center on what will become the compression stroke. Install the camshaft; it should slip easily under the tappets. Rock the crankshaft a degree or two on each side of TDC, alternately engaging the intake and exhaust valves. Timing is correct when freeplay splits evenly between the two valves. If one valve leads the other, reposition the camshaft one tooth from that valve.

Crankshafts that run on plain main bearings extract easily without interference with the camshaft. Timing marks should be clearly visible. Some engines employ the crankshaft gear keyway as one of the marks. Tecumseh-made Craftsman engines with fixed-adjustment carburetors are advanced one tooth for sake of mid-range torque (B of FIG. 6-51). As far as I know, these are the only small engines deliberately mistimed.

Antifriction (ball or tapered roller) bearing cranks can present something of an extraction problem. The top-side bearing rides in a carrier and, because of limited space, the camshaft must be dropped out of position to maneuver the crankshaft throw out of the block. Timing marks on the crankshaft side often take the form of a chamfered tooth, or can be stamped on the counterweight (FIG. 6-52).

Displace the camshaft by driving out the cam axle through the magneto side of the block (FIG. 6-53). Note the expansion plug, which should be oil-proofed with sealant before assembly. Timing might be easier if the associated crankshaft gear tooth is marked with chalk or a crayon. This is particularly true on Briggs & Stratton 30400 and 320400 models.

Figure 6-54 shows inspection points for a Briggs & Stratton plain bearing crankshaft. Other makes do not have the integral point cam, represented by the flat on the magneto end, and there is normally no need to measure journal bearing diameter when antifriction bearings are fitted. These journals do not wear unless the race has spun (in which case the crank might not be salvagable). The crankshaft shown has an inte-

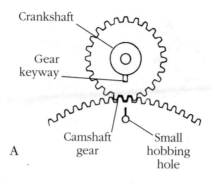

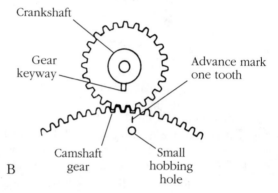

6-51 *When conventional timing marks cannot be found, look for other salient features that could be used as a reference. Nearly every engine built is timed to the marks (A); Tecumseh-built Craftsman engines with suction-lift carburetors are an exception and underscore the need to check timing before disassembly.*

gral gear, which is a bit unusual. Crank and cam drive gears should be renewed as pairs.

Pay special attention to the crankpin. Check for out-of-round, taper, and concentric wear as described under "Connecting Rods." Remember that deep scratches are grounds for rejection. As mentioned in that section, needle bearing crankpins require a glass-smooth surface without the slightest hint of galling or rust pitting. When oversized connecting rod bearing inserts are available, the crankpin can be ground to fit.

Check oil ports for burrs and shavings that could restrict oil flow to the bearings. Crank journals should be lightly polished before assembly with 600 wet-or-dry emery cloth saturated in

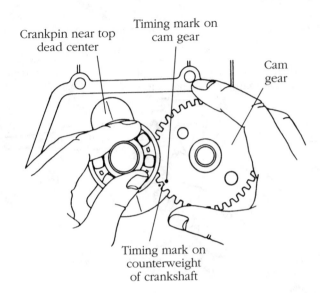

Crankpin near top
dead center

Timing mark on
cam gear

Cam
gear

Timing mark on
counterweight
of crankshaft

6-52 *Axle-supported camshaft on engines with antifriction main bearings must be dropped out of mesh before the crankshaft can be withdrawn. Briggs & Stratton is shown; Clinton is similar.*

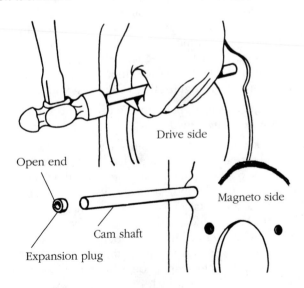

Drive side

Open end

Magneto side

Cam shaft

Expansion plug

6-53 *When fitted, the camshaft axle drives out through an expansion plug. Check axle for wear and coal plug with sealant before assembly.*

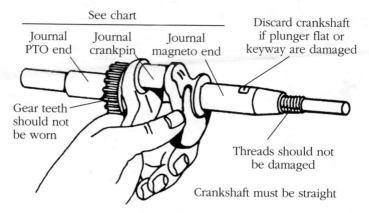

Discard crankshaft if small or out of round

See chart

Journal PTO end Journal crankpin Journal magneto end

Discard crankshaft if plunger flat or keyway are damaged

Gear teeth should not be worn

Threads should not be damaged

Crankshaft must be straight

6-54 *Briggs & Stratton crankshaft inspection procedure applies to other makes as well.*

oil. The best way to do this is to cut a strip of emery paper to journal width, wrap it around the journal, and spin it with a shoelace or leather thong. Remove all traces of abrasive from the crank and, particularly, the oil drillings.

Straightening small engine crankshafts is a touchy subject fraught with legal complications for a mechanic who gets someone hurt. Even so, experienced and patient craftsmen can straighten cranks bent a few thousandths. If you want to pursue the matter, recognize that you are on your own.

Warning: No small engine manufacturer recommends that crankshafts be straightened.

The work requires two machinist's V-blocks, two dial indicators, and a straightening fixture that is usually built around a hydraulic jack. The crank is supported by the blocks at the main bearings and the indicators are positioned near opposite ends of the shaft. Total runout should be no more than 0.001 inch (or 0.002 inch indicated). Using the fixture, the crank is brought into tolerance in small increments with frequent checks. Once the indicators agree, the crank is then sent out for magnetic particle inspection to detect possible cracks. Skipping this final step, which only costs a few dollars at an automotive machine shop, can be disastrous for all concerned.

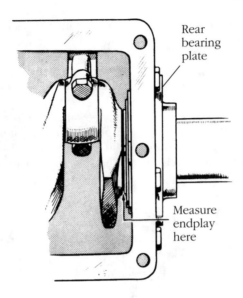

Rear
bearing
plate

Measure
endplay
here

6-55 *Crankshaft endplay, or float, can be measured at any point along the length of the crankshaft with a suitably mounted dial indicator or feeler gauge. Onan Shown.*

Upon assembly, check the crankshaft end play. Depending upon block construction, this check is made internally (FIG. 6-55) or with a dial indicator from outside of the engine. The amount of float is not crucial so long as parts have ample expansion room. Typical specs fall in the 0.003-inch-to-0.005-inch range. Several engine makers supply thicker base or bearing cover gaskets for use when the float has been absorbed by a new crank or flange. A thrust washer—usually placed between crank and PTO main and, occasionally, on the magneto side—compensates for wear.

Camshafts

The camshaft rides on an axle pin (as shown in FIG. 6-53), or else is supported by plain bearings at the magneto side of the block and the topside cover. The latter type can be removed and installed without compressing valve springs if first turned to the timing-mark index position.

Most camshaft failure is obvious: Once the surface hardness goes, the lobes wear round, the gear teeth break, or the gear fragments. A careful mechanic will measure valve lift and bearing clearance. In most cases, cam drive side bearings should be replaced when wear ex-ceeds 0.005 inch or so.

The cam might include a compression release to aid starting (FIG. 6-56). Briggs & Stratton displaces the cam laterally when the starter engages to unseat the exhaust valve. Little can go wrong with this mechanism. Check operation by hand and

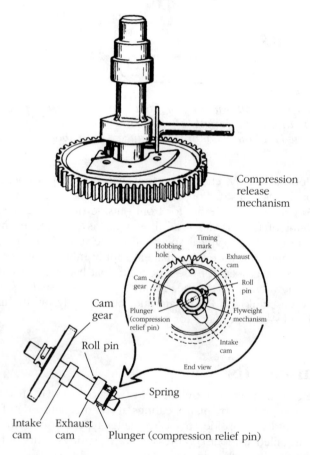

6-56 *Centrifugal compression release should be cycled manually. Tecumseh unit shown is integral with the camshaft.*

look for wear on pivots and weight stops. Clinton and other manufacturers sometimes employ a cam-actuated advance mechanism on engines with side-mounted magnetos. Verify operation by hand. Remove springs only as necessary for replacement.

Main bearings

The crankshaft runs against plain or antifriction bearings or a combination of both types. Plain bearings can be made of brass or are integral with an aluminum block. Antifriction bearings are usually present as ball or roller bearings with inner cones and outer races (cups) to protect both the crankshaft and the castings. Some two-cycle engines use needle bearings riding directly on the crank. Antifriction bearings should be replaced at first sign of roughness and as part of every engine rebuild.

Antifriction bearings

Figure 6-57 shows the more-or-less typical setup using two tapered roller bearings with a washer and shims at the topside to control end play. Check by removing all traces of lubricant from the bearings and spinning the outer races by hand. Roughness or the tumbler-like noise of loose cones means that the bearing should be renewed.

Caution: Do not spin antifriction bearings with compressed air. In addition to damage from water in the air source, the turbine effect will overspeed the bearings.

Remove the bearing from the crankshaft with aid of a bearing splitter (FIG. 6-58). Once drawn in this manner, bearings cannot be reused. The preferred method of installation is to heat the bearing in a container of oil until the oil begins to smoke (a condition that corresponds to a temperature of about 375 degrees F). The bearing should be supported off the bottom of the container with a wire mesh. The more usual method is to press the bearing cold by supporting the crankshaft at the web and applying force to the inner race only. Figure 6-59 shows this operation for Kohler double-press fit bearings. First the bearing is pressed into its cover with the arbor against the outboard race, and then the cover assembly is installed with press force confined to the inner race.

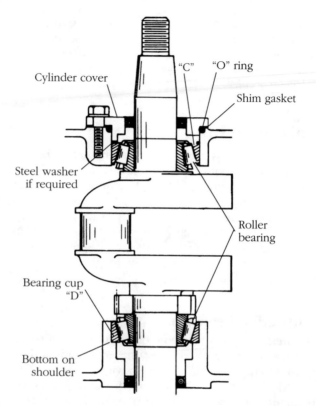

Cylinder cover · "C" · "O" ring · Shim gasket · Steel washer if required · Roller bearing · Bearing cup "D" · Bottom on shoulder

6-57 *Better engines use tapered roller bearings, which absorb thrust as well as radial loads. Shim gasket and optional washer determine crankshaft float.*

Antifriction bearings seat flush against the shoulders provided. Check end play against specification and adjust as necessary with gaskets or shims.

Antifriction bearings are hardware items that can be purchased from bearing-supply houses at some savings over dealer prices. However, be certain that the replacement matches the original in all respects. Unless you have certain information to the contrary, do not specify the standard C1 clearance for bearings with inner races. Ask for the looser C3 or C4 fit, both of which allow room for thermal expansion.

Plain bearings

Determine bearing clearance with inside and outside micrometers and compare with factory specs for the engine in ques-

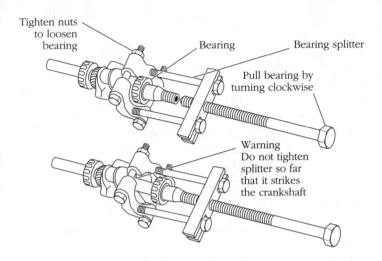

Tighten nuts to loosen bearing

Bearing

Bearing splitter

Pull bearing by turning clockwise

Warning
Do not tighten splitter so far that it strikes the crankshaft

6-58 *Antifriction bearings remain on crankshaft unless they will be replaced.* Clinton

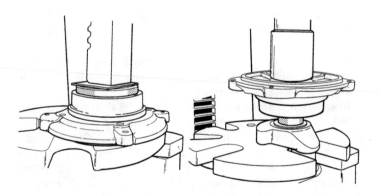

6-59 *Kobler K482 and K532 series engines employ PTO bearings with a double interference fit. The cup (outer race) presses into the bearing cover, and then the inner race, together with the assembled bearings, presses over the crankshaft. A support under the crankshaft web nearest to the arbor isolates the crankpin from bending loads.*

tion. Most are set up with 0.0015 inch new clearance and tolerate some 0.0030 inch before rework.

All engines from major manufacturers can be rebushed, but this is not a do-it-yourself project. The work is best left to a dealer who has access to the necessary reamers, pilots, and drivers.

Thrust bearings

Thrust bearings are normally present as a hardened washer at the top end of the crankshaft. Kohler and other manufacturers sometimes specify a proper babbit-coated or roller thrust bearing. Poorly maintained vertical-shaft engines will develop severe galling at the flange thrust face, which can be corrected by resurfacing the flange or replacing the casting.

Seals

Seals, mounted outboard of the main bearings, contain the oil supply for four-cycle engines and seal crankcase pressure in two-strokes. Seal failure can be recognized by oil leaks at the crankshaft exit points or, on two-cycle engines, by hard starting and chronically lean fuel mixtures. Seals must be replaced to protect the investment in a rebuilt engine.

Install replacement seals with the maker's mark visible and the steep sides of the lip toward the pressure. Lubricate the lip with light grease and, unless the seal is already covered with an elastomer, coat the metal rim outside diameter with gasket sealant. Be careful not to allow the sealant to contaminate the seal lips or the oil return port.

Installation is best done with a factory seal driver that concentrates force on the rim outside diameter. A length of pipe of the appropriate dimension will suffice (FIG. 6-60). Drive the seal to the original depth (usually flush or just under flush), unless the crankshaft seal area is worn. In that case, adjust seal depth to engage an unworn area on the crank, but do not block the oil return port in the process.

The crankshaft must be taped during installation to protect seal lips from burrs, keyway edges, and threads. Celophane tape, because it is relatively thin, works best.

Governor mechanisms

The unit shown in FIG. 6-61 is typical of the crankcase part of most mechanical governor assemblies. Paired flyweights, driven at some multiple of engine speed by the camshaft, pivot outward with increasing force as rpm increases. This motion is

Seal
sleeve
tool

Drive down
with hammer
until seal is
flush with
cover

Oil
seal

Use this method to drive oil seals flush
and square into the seal receptacle

6-60 *Use the correctly sized driver to confine installation stresses to the outer edge of the seal retainer.*

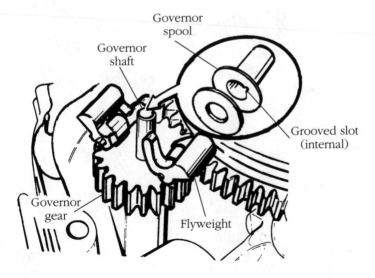

Governor
spool

Governor
shaft

Grooved slot
(internal)

Governor
gear

Flyweight

6-61 *Typical centrifugal governor. Flyweights react against a plastic spool.*

translated into vertical movement at the spool and appears as a restoring force on the carburetor throttle linkage. Work the mechanism by hand, checking for ease of operation and obvious wear. The governor shaft presses into the block or flange casting; in event of replacement, it must be secured with Loctite bearing mount and installed to the prescribed height.

Oiling systems

Two-cycle engines are, of course, lubricated by oil mixed with the fuel. Bearings can be vented with ports or milled slots to encourage oil migration.

Four-cycle systems are more complex and worthy of special attention. Indeed, the best mechanics will not release an engine unless all circuits have been traced, cleaned with rifle brushes, and buttoned up with new expansion plugs.

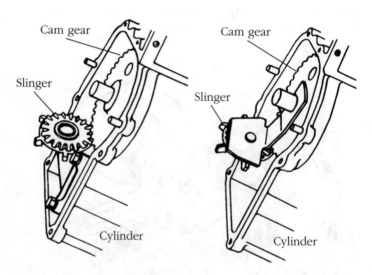

6-62 *Early (left) and late-production Briggs & Stratton slingers. The later variant might incorporate governor weights and a wave washer between the bracket and flange. Bracket should be replaced when camshaft hole has worn to 0.49 inch or larger.*

Any of the three oiling systems are used. Most small side-valve engines depend upon a *splash system*, in which crankcase oil is agitated by a dipper mounted on the connecting rod cap or camshaft gear. Briggs & Stratton engines in this class employ a camshaft-driven slinger (FIG. 6-62). Other than thoroughly cleaning the inside of the crankcase, checking dipper orientation (another reason to make certain the rod cap is on right), and inspecting the slinger for wear, no special maintenance procedures are required.

Semi-pressure systems combine splash with positive feed to some bearings. The Tecumseh system, used on vertical crankshaft engines, is fairly typical of the breed (FIG. 6-63). A small plunger-type pump (FIG. 6-64), driven by the camshaft, draws oil from a port on the cam during the pump intake stroke. As the plunger telescopes closed, a second port on the camshaft hub aligns with the pump barrel and oil is forced through the

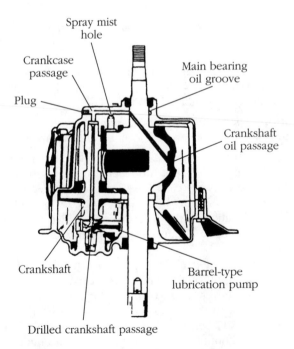

6-63 *Tecumseh system pumps oil to the upper main bearing and crankpin on vertical-shaft engines. Remaining parts lubricate by splash.*

hollow camshaft to a passage on the magneto side of the block. Cross drillings in the camshaft provide lubrication to the bearings. Once in the block passage, the oil flows around a pressure-relief valve (set to open at 7 psi) and into the upper main bearing well. Most models feature a crankshaft drilling to provide oil to the crankpin.

Flat

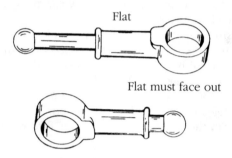

Flat must face out

6-64 *Plunger pump drives off a camshaft eccentric.*

Blow out the passages with air and check the pump for scores and obvious wear. Replace the pump plunger and barrel as a matched assembly.

Caution: The pump must be assembled with the flat side out and primed with clean motor oil before startup.

Some Tecumseh engines use an Eaton-type oil pump, recognizable by its star-shaped impeller. Check for scuffing on the impeller and pump case ID. Clearance between impeller and pump cover should be gasketed to 0.006–0.007 inch. Except for relocation of the pressure relief valve in the flange between the pump and camshaft, oil circuitry is as previously discussed.

Full-pressure systems deliver pressurized oil to all crucial bearing surfaces, although some parts receive lubrication from oil thrown off the crankpin (cylinder bore, cam gear), or by oil flowing back to the sump (valve guides). The Kohler system, used on KT17 Series II and KT19 Series II engines, is typical of most—although circuitry varies between engine makes and models (FIG. 6-65). A conventional gear-type pump supplies oil to the topside main, No. 1 crankpin, and to the camshaft, which serves as a gallery to bring oil to the magneto-side main bearing and No. 2 crankpin. A pressure relief valve under the topside main bearing carrier limits pressure to 50 psi to prevent bearing erosion. Provision for an oil pressure sender is by way of a ⅟₁₆-inch NPTF plug on the topside of the crankcase.

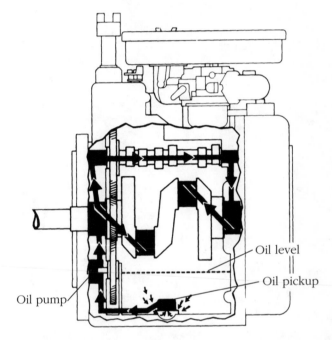

6-65 *Kohler full-pressure system utilizes a gear-driven pump and hollow camshaft.*

Index